Risk and Society: The Interaction of Science, Technology and Public Policy

Technology, Risk, and Society

An International Series in Risk Analysis

VOLUME 6

The titles published in this series are listed at the end of this volume.

RISK AND SOCIETY: THE INTERACTION OF SCIENCE, TECHNOLOGY AND PUBLIC POLICY

Edited by

MARVIN WATERSTONE

SPRINGER-SCIENCE+BUSINESS MEDIA, B.V.

Library of Congress Cataloging-in-Publication Data

Risk and society : the interaction of science, technology, and public policy / edited by Marvin Waterstone.
p. cm. -- (Technology, risk, and society ; v. 6)
Includes index.
ISBN 978-94-010-5613-7 ISBN 978-94-011-3634-1 (eBook)
DOI 10.1007/978-94-011-3634-1
1. Technology--Risk assesment--Social aspects. 2. Science--Social aspects. I. Waterstone, Marvin. II. Series.
T174.5.R555 1991
363.1--dc20 91-806

ISBN 978-94-010-5613-7

Printed on acid-free paper

Originally published by Kluwer Academic Publishers in 1992
Softcover reprint of the hardcover 1st edition 1992

Table of contents

List of Tables and Figures

Tables

Chapter

Figures

Contributors

JAMES K. ASSELSTINE is an electric utility analyst with Shearson Lehman Hutton. Prior to this position Mr. Asselstine held a similar position with the Donaldson, Lufkin and Jenrette Securities Corporation. From May, 1982 until July, 1987, Mr. Asselstine served as a Commissioner with the U.S. Nuclear Regulatory Commission. Prior to his appointment as an NRC Commissioner, Mr. Asselstine served as Associate Counsel on the staff of the Senate Committee on Environment and Public Works, and in that position served as a codirector of the Committee's investigation of the Three Mile Island accident. Mr. Asselstine is an attorney by training, and received his law degree from the University of Virginia.

DAVID S. BARON graduated Phi Beta Kappa from Johns Hopkins University and Cum Laude from Cornell Law School, where he was Article Editor for the Cornell Law Review. He clerked for Judge Anthony Celebrezze of the U.S. Court of Appeals for the Sixth Circuit. Following that position, Mr. Baron conducted environmental enforcement actions as an Assistant Attorney General for the State of Arizona. Since 1981, Mr. Baron has conducted environmental litigation and advocacy for the Arizona Center for Law in the Public Interest on issues including water pollution, hazardous wastes, air quality and protection of public lands. Mr. Baron has also studied environmental issues in Europe as a fellow of the German Marshall Fund of the U.S.

THERESA A. CULLEN, M.D. is currently working as a family practice resident at the University Medical Center in Tucson, Arizona. Prior to her current position, Dr. Cullen was a physician for the Indian Health Service at Sells, Arizona. As maternal child health coordinator for the Sells Service Unit, she helped implement a comprehensive prenatal education program, a clamydia screening program, and a car seat loaner program. Before taking the position at Sells, Dr. Cullen served as a staff physician at the San Carlos (Arizona) Service Unit of the Indian Health Service. During that time she coordinated emergency medical services and developed standard prenatal care policies. Dr. Cullen also has a strong interest and background in community organization, having been an organizer for the United Farm Workers, and a community development worker for the Cree Nation.

SUSANNA EDEN is a Senior Research Specialist at the Water Resources Research Center at the University of Arizona. Before joining the Center, she was a Graduate Research Assistant at the University and received an M.S. from the Department of Hydrology and Water Resources in water resources–planning and policy. Her publications include papers on water quality management, integrated water management, and negotiation of water disputes.

MORRIS FARR has a bachelors degree in physics from Rice University, and M.S. and Ph.D. degrees in nuclear science from the University of Michigan. His research areas include controlled thermonuclear fusion and reactor physics. He has taught at the University of Arizona since 1969 in the Department of Nuclear and Energy Engineering. He has also performed research at Oak Ridge National Laboratory, Lawrence Livermore Laboratory and Los Alamos National Laboratory in this country, and at Culham Laboratory in England. Dr. Farr also served three terms in the Arizona State Senate where he rose to the rank of Democratic whip. Dr. Farr is currently an Assistant Dean of the College of Engineering and Mines at the University of Arizona.

HELEN INGRAM is Director of the Udall Center for Studies in Public Policy at the University of Arizona, and a Professor with a joint appointment in the Department of Political Science and the School of Public Administration and Policy. Her research has been in the areas of water resources, natural resources and the environment, U.S.-Mexico border problems and policy implementation. She has written extensively on southwestern water issues, and is the Review Editor for the Journal of the American Political Science Association.

ALBERT JONSEN is Professor and Chair of the Department of Medical History and Ethics in the School of Medicine of the University of Washington. He came to the University of Washington from the University of California, San Francisco where he had been Chief of the Division of Medical Ethics since 1972. Prior to that he was President of the University of San Francisco, where he taught in the Departments of Philosophy and Theology. He received his doctorate from the Department of Religious Studies at Yale University in 1967. Dr. Jonsen was elected a member of the Institute of Medicine of the National Academy of Sciences in 1980. He served as Commissioner on the National Commission for the Protection of Human Subjects of Biomedical and Behavioral Research from 1974-1978, and on The President's Commission for the Study of Ethical Problems in Medicine from 1979-1982. Dr. Jonsen has written extensively on the topics of medicine, health care and ethics.

WENDY LAIRD is a Research Fellow at World Wildlife Fund and The Conservation Foundation where she is staff to the National Commission on the Environment. Previously, she was Program Coordinator for Linda Laird & Associates (an architectural historic preservation consulting firm); and a Graduate Teaching and Research Assistant at the University of Arizona where she co-authored several articles on agenda-setting, received a M.P.A., and was inducted into Pi Alpha Alpha. She was awarded a Presidential Management Internship in 1990.

MICHAEL D. LEBOWITZ is Professor of Internal Medicine at the University of Arizona in Tucson. He is also Associate Director of the Division of Respiratory Sciences in charge of Environmental and Epidemiological Programs. He was senior editor of the OMS/WHO Guidelines on Studies in Environmental Epidemiology (Environmental Health Criteria # 27), was Co-Chair of the National Academy of Science/National Research Council's Committee on Indoor Pollutants, and was co-author of the American Thoracic Society Statement on the Health Effects of Air Pollutants. He has served on committees for the Environmental Protection Agency, the World Health Organization, the National Academy of Sciences and the National Institutes of Health. He presently serves on the WHO/EURO Working Committee on Indoor Air Quality, and the EPA Science Advisory Committee Sub-Committee on Strategic and Long-Range Planning. He has written approximately 300 articles, abstracts, chapters and books, and serves on the editorial boards of several technical journals.

DEBORAH MATHIEU received a Master's degree in Religious Ethics from Yale University Divinity School and a Ph.D. degree in philosophical ethics from Georgetown University. She is currently an Assistant Professor of Political Science and Philosophy at the University of Arizona. Dr. Mathieu specializes in medical ethics and health policy, and her published work in these fields includes books on organ substitution technology and emergency medicine, as well as journal articles on related topics.

RONALD D. MILO is Professor of Philosophy at the University of Arizona, where he has taught since receiving his Ph.D. from the University of Washington. His research areas include moral, social and political philosophy, with a specialty in ethical theory. He has published several books and a number of articles in those areas. Currently he is at work on a book about the relationships between morality and social conventions.

H. BRINTON MILWARD is Director of the School of Public Administration and Policy, and Associate Dean of the College of Business and Public Administration at the University of Arizona. He is also an Associate Professor of Management and Policy and Public Administration. He received his doctorate from the Ohio State University. He has published widely and is a nationally-known scholar in public administration who serves on the editorial boards of several of the leading journals in the field. He has worked and consulted for the states of Ohio, Kansas, New York, and Kentucky, as well as for the U.S. government, cities, foundations and national associations. His recent research has focused on industrial development and the impact of Japanese investment in the U.S., and on the relationship between ideas and interests as they relate to setting the national policy agenda.

NORMAN J. PETERSEN is Chief, Office of Risk Assessment and Investigation, Division of Disease Prevention, Arizona Department of Health Services. In this position he serves as the State Environmental Epidemiologist and manages programs involving standard setting, studies of environmentally-provoked diseases, pesticide poisoning prevention and investigations of toxic substance exposure. Mr. Petersen received a B.S. degree in civil engineering from the University of Wisconsin, and a S.M. degree in radiologic health from Harvard University. Before joining the Arizona Department of Health Services in 1983, he served as a commissioned officer in the U.S. Public Health Service for 28 years. During that period, Mr. Petersen was involved in a wide variety of research assignments as an engineer-epidemiologist, and authored or co-authored numerous papers concerning the role of the environment in public health.

WILLIAM D. ROWE is currently President of Rowe Research and Engineering Associates, Inc. in Alexandria, Virginia. Prior to that he was a Professor of risk and decision sciences at American University from 1978 to 1987. Before joining American University, Dr. Rowe was the Deputy Assistant Administrator for Radiation Programs at the U.S. Environmental Protection Agency for eight years. He received a doctorate in operations research from American University in 1973. Dr. Rowe has over 35 years experience in industry, government and academia. He has published six books in the area of risk analysis, has published over 200 papers and holds seven patents. His present expertise is directed at the whole field of risk analysis, particularly in the environmental area. Dr. Rowe has an extensive background in computers and simulation and modelling, and has been honored as a pioneer of the Society for Computer Simulation.

MILTON RUSSELL at present holds a joint appointment with the University of Tennessee, Knoxville, and Oak Ridge National Laboratory. At UTK he is a Professor of Economics and Senior Fellow in both the Waste Management Research and Education Institute and the Energy, Environment and Resources Center. Before coming to Tennessee, he served as an Associate Administrator of the U.S. Environmental Protection Agency under both William D. Ruckleshaus and Lee M. Thomas, directing the Agency's policy, planning, regulatory development and evaluation functions. At EPA he was particularly charged with helping to develop systems and procedures for implementing the risk-based approach to environmental management. While he has specialized in environmental policy since 1983, before that Dr. Russell's research and work focused on energy policy. He was Senior Fellow and Director of Resources for the Future's Center for Energy Policy Research. From 1974-1976, he also served as the Senior Staff Economist covering energy policy at the President's Council of Economic Advisors.

MARVIN WATERSTONE is Associate Professor of Geography and Regional Development at the University of Arizona, and Associate Director of the UA Water Resources Research Center. He received an M.A. Degree in Geography from the University of Colorado, and a Ph.D. in Geography from Rutgers University in 1983. He has worked in the areas of environmental policy, natural resources, risks and hazards for over 15 years. His current work focuses on the social generation of risks and hazards, and includes research on institutional innovation in resource management, global environmental change, Amazon deforestation, the impacts of military spending on local and regional economies, and the interactions of science, technology and society. Prior to joining the University of Arizona, Dr. Waterstone taught at Kent State University and San Diego State University. He has been a consultant for city, state and federal agencies as well as the private sector. He has published widely in the areas of risk and hazard management, natural resource and water policy, and aspects of science and society.

Preface

Life in the last quarter of the twentieth century presents a baffling array of complex issues. The benefits of technology are arrayed against the risks and hazards of those same technological marvels (frequently, though not always, arising as side effects or by-products). This confrontation poses very difficult choices for individuals as well as for those charged with making public policy.

Some of the most challenging of these issues result because of the ability of technological innovation and deployment to outpace the capacity of institutions to assess and evaluate implications. In many areas, the rate of technological advance has now far outstripped the capabilities of institutional monitoring and control.

While there are many instances in which technological advance occurs without adverse consequences (and in fact, yields tremendous benefits), frequently the advent of a major innovation brings a wide array of unforeseen and (to some) undesirable effects. This problem is exacerbated as the interval between the initial development of a technology and its deployment is shortened, since the opportunity for cautious appraisal is decreased.

One major impetus for this phenomenon is the blurring of roles that have been separated more distinctly in the past; that is, between the scientist/technologist on the one hand, and the entrepreneur on the other. Within the past forty years this separation has become increasingly difficult to discern, and in fact, in many instances today is non-existent. The individuals developing technological innovations are the same people who are motivated to deploy and market the new technology or product, and in most cases, as rapidly as possible. A critical layer of review and evaluation has therefore been largely eliminated.

One of the vital considerations which was possible to include in this review period (although this was certainly not always the case) was the incorporation of public values. The advantages of a new technology could be examined in the context of social preferences, and could be weighed against alternatives and against potential adverse effects or risks. Today the process is being short-circuited to a greater degree than ever before. Consequently, the types of risks and hazards, as well as the ways in which they arise are changing, and, in many cases, are overwhelming our ability as a society to cope effectively.

The litany of such issues is certainly long, but prominent among them are the following: surrogate motherhood; in vitro fertilization and embryo transfer; heroic medical interventions (and the concomitant ethical issue of the allocation of scarce medical resources); recombinant-DNA research, gene splicing and other biotechnologies; the Strategic Defense Initiative; nuclear power and nuclear waste disposal; natural hazards; dam safety; and setting public health standards (e.g., for drinking water, air quality or radiation exposures).

All of the foregoing have at least three major elements in common. First, all are generated as the result of advances in technology, either as an initiating agent (i.e., a particular technology is the direct cause of a problem), or as a remedial agent (i.e., a technological approach has been deployed to address a pre-existing problem). Second, all involve risk and uncertainty. And third, all require an assessment of profound human values and social preferences for resolution.

Commonly such issues are dealt with in the public policy arena, and necessitate a high level of interaction between scientists and policy makers. This science/policy interface is one of the areas where science, technology and human values intersect most overtly, although the fit is often an uneasy one. Policy makers are required to evaluate not only the technical and scientific merits of proposed courses of action, but also the ethical, moral and social (not to mention, political) costs and benefits.

In order to accomplish this, substantial communication must occur between scientists/technologists and policy makers (who frequently, though not always, have less grounding in scientific/technical matters). Dialogue must also take place between policy makers and their constituents so that policies can best reflect the attitudes, beliefs and values of the citizenry.

Meaningful participation in this process on the part of the public also demands a significant amount of information not only about specific issues, but also about the nature of risk and uncertainty in general. The public at large, as well as their appointed and elected officials, must understand the substantial levels of uncertainty inherent in scientific judgments. They must also be alerted to the types and magnitudes of risks involved in pursuing particular future paths.

Scientists in general, and those in universities in particular, have a significant responsibility for sharing their knowledge with the public and with policy makers regarding developing technologies, their potential benefits and risks, and the uncertainties involved in scientific judgment. Informed public policy can emerge only through a clear and comprehensive explication of all of the relevant issues, insofar as this is possible.

However, this communication process must also be viewed as multidirectional. Scientists, technologists and policy makers must also be prepared to listen to public concerns over technological change, and must be sensitive to the dimensions of risk issues that are relevant to lay people. A mismatch between expert and lay opinions on risk does not automatically define the latter as irrational. In many cases, members of the public place a higher priority (and value) on different aspects of risk than do the body of "experts."

This volume is one step in such an information process. The book is intended to explore the notion of risk as a constructed phenomenon. It deliberately moves away from the idea that most risks are the result of accidents or processes that are out of human control.

The discussion begins with an overview chapter which examines the ways in which risks and hazards have been examined in the past, and then moves on to examine this alternative view of risks, particularly as they arise out of ongoing social processes and institutions in society. The book then moves to four topical areas, each explored by three authors with differing perspectives.

The first of these sections is intended as a general assessment of science-based public policy making for risky issues, and the role of risk assessment in such processes. Part two turns to a discussion of health-based risks. However, the focus here is not the usual one on environmental pollutants or other common topics in health risk discussions. In this section, in keeping with the theme of risks arising out of ongoing social relationships, the authors explore the risks associated with the allocation of scarce medical resources. Part three moves to a more familiar technological risk topic, nuclear power. The final section of the book examines the general policy problem of setting standards for public health and safety. The section utilizes the problem of air pollution as a way to explore this complex set of issues.

None of the authors (with the exception, perhaps, of the editor) can be considered as part of the mainstream of researchers on risks and hazards. Yet each offers important insights on the ways in which risk issues arise, mature and have profound impacts upon our society. It is my hope that by making the discourse about risks and hazards more ecumenical, we can achieve a better understanding of the nature of these so-called "side effects" of our changing society.

Acknowledgements

This volume emerged out of a symposium of the same title held at the University of Arizona over the Fall 1988 and Spring 1989 semesters. The symposium was supported by the efforts of several, enlightened units and administrators at the University of Arizona. For their assistance, I want to thank Dean James Dalen of the College of Medicine, Dean Edgar McCullough of the Faculty of Science, Dean Lee Sigelman of the Faculty of Social and Behavioral Sciences (and at the time of the symposium, Acting Director of the Udall Center for Studies in Public Policy), and Dean Ernest Smerdon of the College of Engineering and Mines. Their support made the symposium, and ultimately this volume possible.

I would like to also acknowledge the able support of several staff members of the Water Resources Research Center at the University of Arizona. These include Pamela Hathaway, who served early on to coordinate the many details of the symposium; Susanna Eden, who took over from Ms. Hathaway, and saw the project through to completion. I would also like to thank Ana Rodriguez for taking care of the many accounting details associated with the project. Finally, I owe my thanks to Yvette Semler, who cheerfully and capably handled all of the numerous word processing duties.

The Social Genesis of Risks and Hazards

MARVIN WATERSTONE

Natural, technological and social hazards form a central part of modern life. The variety, magnitude and extent of such hazards is enormous, ranging from everyday risks (e.g., automobile accidents, unsafe drinking water and the like) to rare, but potentially cataclysmic events (e.g., large earthquakes or nuclear holocaust). In the past, most hazards research has focused largely on delineation of hazard occurrence (location, magnitude, extent, periodicity), on the underlying physical causes of hazards, and on human strategies for managing or reducing hazard consequences.

Relatively little research has been undertaken to identify the societal phenomena which give rise to hazardous conditions (for examples which move in this direction, see Bogard, 1988; George, 1982; Hewitt, 1983; Johnson and Covello, 1987; Palm, 1990; Perrow, 1984; Sjoberg, 1987). Why do hazards and risks arise in a particular manner? Are hazardous conditions produced differently from one cultural setting to another? If so, how and why? Is the way in which hazards are being generated undergoing change as we move further into the technological age? If so, what implications do these changes pose for our institutional capacity to manage hazards and risks? This chapter seeks to establish some structure for such questions, and to provide some tentative conclusions regarding the contextual factors that generate hazards and their accompanying risks.

The chapter begins with a brief examination of what might be thought of as the predominant praxis of past and current research on risks and hazards. In that section, I make the case that while much of this research may be useful for addressing one set of hazard problems, it has failed to address a much broader (and more significant) set of risk and hazard issues. The second section of the chapter focuses on this broader set of issues and documents the need for risk research to come to grips with these types of threats. The final section of the chapter suggests an approach for investigating such risks and hazards and applies the approach to an examination of the potential risks associated with emerging biotechnologies.

M. Waterstone (ed.), Risk and Society: The Interaction of Science, Technology and Public Policy, 1–12.

Overview of Predominant Hazards Paradigm

There has been much written of late which critiques the predominant approaches of hazards research. It has been suggested that this research has been largely atheoretical; has taken a mechanistic, deterministic view of events and behavior; has been scientistic and technocratic; has largely downplayed, if not ignored, the role of social and economic factors in affecting risk; and has represented an ideology of the status quo (for examples, see Brown, 1977; Cliffe, 1974; Hewitt, 1983; Kirby, 1990; O'Keefe, Westgate and Wisner, 1976; Sewell and Foster, 1976; Tinker, 1984; Torry, 1978a, 1978b; Wisner, O'Keefe and Westgate, 1976).

While such arguments have much merit, it is not the central purpose here to dwell on these critiques. Rather my intention is to take a slightly different point of departure, and focus on the subject matter of most risk and hazard research rather than directly on its methodological or ideological shortcomings. To be sure, ideology and method are intrinsically tied to subject matter issues, and these connections are explored to the degree they are relevant to present purposes.

By and large, most research on risks and hazards (past and current) has tended to focus on discrete, highly identifiable threats–geophysical events (e.g., a flood or earthquake) or isolated technological hazards (e.g., a nuclear plant incident or a hazardous material spill). Also most commonly, the focus has been on extraordinary, relatively infrequent events, sometimes (particularly in the case of technological hazards) referred to as accidents. A concurrent concern, given this orientation toward extraordinary events (i.e., seen as largely uncontrollable), has been a focus on human coping strategies (either by individuals or groups). Such coping has most often been described in terms of a more or less rational choice process which includes an information gathering phase, an assessment of costs and benefits, and finally a selection of a strategy. This focus on rational (even boundedly rational) coping has also been critiqued in the literature (e.g., see Waterstone, 1989; Waterstone and Lord, 1989; Watts, 1983), and those arguments need not be replicated here.

While such events are significant in terms of inflicting such annual costs as lives lost, human injuries, and environmental and property damage, they are far outshadowed in these dimensions by what might be considered the pervasive risks and hazards which people face on a continual basis. These latter risks do not stem from extraordinary events, are not thrust into the spotlight frequently by media exposure, and do not often mobilize the technological response apparatus brought into place for the extraordinary events which are the focus of most risk and hazards research.

There is one other significant difference between such pervasive risks and the discrete events of hazards research. For the most part, people do not "take" these pervasive risks, but are exposed to them, usually without their knowledge or consent. These are the risks which arise from the "normal," day-to-day functioning of society. The kinds of risks which arise most frequently from the interactions of science, technology and society (i.e., the subject of this volume) are of this type.

I am arguing here that risk and hazards research has largely failed to examine these types of risks, and through this failure has neglected the most significant source of threat to people and their environment. The rest of this chapter is devoted to documenting this type of pervasive risk, and to exploring the research implications of reorienting the definition of risk away from discrete, extraordinary events to one of pervasiveness and ordinariness.

Before proceeding, it is critical to state that this distinction between extraordinary and pervasive risks is made primarily for analytic convenience. It should be noted that many of the so-called extraordinary risks themselves (or at least significant components of them) arise out of the basic ways in which social institutions are constructed, rather than as uncontrolled (or uncontrollable), accidental events. The reader should bear this caution in mind for the discussion that follows.

What Is Meant by Pervasive Risk?

In order to indicate the nature of these pervasive risks, and at the same time, begin to document the ways in which such risks are inextricably tied to the usual functioning of society, I identify four increasingly general case areas. These are only meant to be illustrative, and in no way purport to be comprehensive. Each of these four case areas focuses on different population segments. However, each indicates the ways in which social institutions are capable (either deliberately or through malign neglect) of subjecting people to risks. Each case area also illustrates the enormous degree to which people are exposed to risks without their knowledge or consent. Finally this documentation of everyday, pervasive risk will begin to frame the argument for a new approach to risk and hazard reduction which will be amplified in the last section of the chapter.

Risk Exposure of Captive Populations

Over time, there have been several attempts to document the degree to which selected, captive segments of the population have been exposed to risks involuntarily. Most often such expositions have dealt with prison populations or members of the armed services.

One recent example of the willful exposure of service personnel to unwarranted risks is provided by Uhl and Ensign (1980). This account focuses on the exposure of GIs to nuclear radiation (during atomic testing) and deadly herbicides (including Agent Orange). Clearly these are risks over and above the obvious risks imposed by combat. The authors indicate, for example, that over 300,000 GIs were exposed to ionizing radiation from the atmospheric testing of nuclear devices between 1945 and 1961. While it is not possible to state the precise numbers of GIs who were exposed to defoliants in the Vietnam war, estimates run to the tens of thousands. The original lawsuit on behalf of exposed veterans was a class action which asked the court to certify as a class all 2.8 million veterans who served in Vietnam.

The book provides graphic documentation of the extent to which societies frequently are capable of imposing risks when the ends (e.g., national security, military necessity) are perceived as justifying the means. However, those GIs exposed certainly had little say in the matter, and the accounts raise very troubling questions about the nature of risk and possible response.

Occupational Risks

By this point the literature on occupational health and safety is voluminous and of long-standing (e.g., Ashford, 1976; Nelkin and Brown, 1984; Page and O'Brien, 1973; Scott, 1974; Stellman and Daum, 1973; Wallick, 1972). However, this is another area which has received relatively little attention from the risk and hazards research community. Workers around the world are exposed daily to noise, dust, fumes, vapors, gases, extremes of heat and cold, vibration, radiation, poor lighting and ventilation, untested chemicals, and psychological stress. It is rather easy to bring to mind some of the more spectacular incidents such as the estimated 150,000 miners who developed Black Lung, the 3.5 million workers exposed to asbestos (who now have a lung cancer rate 5 times the average for the general population), the problems with beryllium, vinyl chloride, and many others.

However, many more workers are affected by the less spectacular, but no less deadly effects of daily exposures to workplace risks. There are tens of thousands of fatal job-related injuries each year, as well as several million annual sub-lethal injuries. These figures do not include problems of occupational disease, which is still a high point of contention between management, labor and government. However, as long ago as 1972, the President's Report on Occupational Safety and Health made a conservative estimate that "there may be as many as 100,000 deaths per year from occupationally-caused diseases, and at least 390,000 new cases of disabling occupational disease each year." These numbers have been on the rise since that time despite the enactment of the OSHAct in 1970. Statistics for other parts of the world, particularly from developing nations are sparse and often unreliable, but reveal a similar pattern of high risk.

Such risks are often characterized as voluntary, and in some narrow sense that might be appropriate. To the extent that workers take such risks knowingly (which is most often not the case, despite new worker right-to-know mandates in the U.S. and elsewhere), and to the extent that workers have a choice to move to other jobs, occupational risks might be said to be voluntary. However, given the almost universal lack of such knowledge and/or choice, these risks are more properly classified as involuntary.

Risks to Vulnerable Segments of the Population

To pursue the issue of voluntary vs. involuntary assumption of risk just a bit further, it is worth examining several segments of the population for whom everyday phenomena present special threats. In this category it is possible to include fetuses, newborns and young children, the elderly, and other individuals who are particularly susceptible to risks because of pre-disposing health conditions (e.g., people with upper respiratory problems, heart disease, etc.). In this category one might also consider inter-generational risks (i.e., beyond the next immediate generation).

To illustrate this category briefly, let me discuss the risks imposed on fetuses, newborns and young children. For these individuals, very slight environmental poisoning during gestation and early childhood can result in devastating effects. These include both birth defects and chronic health problems later in life if the individual survives. Fetuses and very young children are much more sensitive to insult than are adults (Norwood, 1980).

As with occupational risks, several graphic examples come immediately to mind in this case. Some, for example, the diethylstilbestrol (or DES) case, provided startling new findings. The discovery that women who had taken this synthetic hormone during their pregnancies had given birth to daughters who later developed genital cancers, provided the first real evidence that cancers could be induced in the womb. Before the DES case, it was not believed that chemicals could produce such transplacental effects. Now, more than 50 substances have been found which can be considered transplacental carcinogens (Norwood, 1980).

Other well-known examples include the effects of lead on brain development, the thalidomide tragedy of the late 1950's (in which mothers had been prescribed this drug as a tranquilizer, and subsequently thousands delivered children with severe birth defects), and the relationship between exposure to organic chemicals and spontaneous miscarriages. [Some research (Norwood, 1980) indicates that the rate of such spontaneous loss may be nearly 45% of all conceptuses. However, it is difficult to calculate the precise number of spontaneous abortions in a population, since many conceptuses do not survive the few weeks necessary for a woman to become aware of the pregnancy.]

In addition to these well-known problems, many other risks confront fetuses and the young. Many of these problems do not show up until later in life, when they are manifested as chronic health difficulties whose etiologies are then difficult to trace. Some trends have been identified, however. For example, some years ago one study indicated that "in much of the advanced world, respiratory disease, which is closely linked with some air pollutants, now takes second place only to cancer as a disease killer of children under the age of fifteen" (Hunt and Cross, 1975).

Finally, there are those childhood difficulties that are less obvious than heart problems, visible birth defects, and respiratory ailments. These include children who are unusually nervous, and who tend to be classified as hyperactive, minimally brain damaged, or learning impaired. While there are a wide variety of possible causes for such problems, increasing evidence indicates that a significant portion of the blame can be placed on environmental agents. Several that have been suggested in addition to lead are: cigarettes and alcohol used by expectant parents, copper, synthetic food additives, trace metals, and certain types of fluorescent lighting (recall fluoroscopes in shoe stores).

Analogous risks can be identified easily for other vulnerable population segments. The risks in these cases pose less significant (although not always clearly delineated) problems for the general population, but those members of vulnerable groups often have little knowledge or choice regarding their exposure.

General Population Risks

In this last category, I simply want to present a partial litany of pervasive risks to which people in the late 20th century are exposed routinely. Again the list is not meant to be exhaustive, only illustrative. Such risks include widespread contaminated air, water and soil; food additives; exposure to ionizing radiation (gamma rays, microwaves, Xrays, power lines); global-scale biophysical changes (e.g., climate change, or ozone depletion); social and psychological stresses (e.g., crime, job insecurity, financial pressures, fear of nuclear holocaust); and many others.

Hazards research, insofar as it has paid any attention to such risks at all, for the most part has only attempted to delineate the extent and magnitude of such problems. There has been little attempt by researchers to identify and assess the genesis of such risks, or to put them into a more meaningful social context. In the final section, the chapter turns to these issues.

Research on Pervasive Risks

So far this chapter has endeavored to make the following points: first, that most past and current hazards research focuses on discrete, extraordinary events; and second, that while this work has often (although not always or consistently) been useful in ameliorating such hazards, it has neglected a more significant problem set, namely the more pervasive, ordinary hazards which confront people on a daily basis. If risk and hazards research were to turn to this second set of problems, how might research be structured? I want to make the case here that the usual approaches of such research will not suffice.

First let me offer a basic assumption: it is a social goal to reduce such pervasive risks to an acceptable level. (Clearly this finesses such critical questions as whether this is a social goal, what constitutes a society in this context, what is an acceptable level, and who decides all of these questions.)

However, if this assumption is valid, there are three possible means for achieving this goal and they are represented in the following relationship:

↑ Risk Response – ↓ Source of Risk

More specifically, one can reduce risks (i.e., the potential for harm) by: 1) increasing the ability of individuals or groups to cope; 2) decreasing the source of risks; or 3) some combination of these. Traditional risk and hazards research has focused heavily on the left side of this relationship (e.g., through the extensive work on choice of adjustments and so forth), although there has been some attention paid to understanding the proximal, as opposed to the more fundamental, underlying sources of risk, particularly for geophysical hazards and for technological accidents.

For the kinds of pervasive risk I have identified, I want to make the case that a focus on risk response will not be useful. There are two main, correlated reasons for this. First, as I described most of these risks are imposed rather than taken. The implication is that risk bearers have very little knowledge of the hazards they confront, clearly a prerequisite for purposive coping action. The second reason gets back to the matter of choice. Most of the pervasive risks I have identified are largely unescapable, and therefore no individual coping actions will suffice. How does one avoid global warming, ozone depletion (or pollution), or DDT and PCBs in the environment through coping strategies?

Going back to our risk reducing strategies then, if increasing response is unlikely to prove very fruitful, one is left with the option of decreasing risk at the source. What does this mean for the types of pervasive risk I have been describing? In essence it means examining those sets of societal and structural relationships which give rise to unacceptable risk in particular ways. Questions here include the role of the state in managing risks, private vs. public control of new technologies, the role of expertise and science, and myriad others.

There are many ways in which such issues could be examined by hazards researchers. In this chapter I want to suggest the outlines of one such approach. If one begins to think of risk genesis from a social/structural perspective, one question that comes to mind immediately is: do different sets of societal or cultural relationships give rise to risks of different types and in different ways? I am suggesting that this issue is amenable to a cross-cultural examination that would prove useful in answering a variety of important questions.

Certainly there are many difficulties in undertaking such a series of studies. One is the unit of analysis issue (i.e., what constitutes a society or culture in this sense). A second is the identification of the critical set of relationships which would have to be examined (e.g., the relationship between government and industry, between government and citizens, between science and industry, between multinationals and governments, and so forth). However, it might be possible to evaluate the role of such relationships by undertaking a case study of one risk vector and tracing the effects of different social/structural relationships to different absolute levels of risk produced (leaving aside the risk measurement issue for the moment). To illustrate how this approach might be applied, I want to examine the social genesis of risk and hazard in one particular case: biotechnology. In this preliminary exploration I have only sketched out the relevant social relations for the U.S. A more complete analysis would place these comments in a comparative, cross-cultural framework to identify salient similarities and differences in institutional characteristics.

The Social Relations of Biotechnology

In order to understand the ways in which social relationships produce pervasive risk, I want to place the emerging issue of biotechnology in its broader social context. To begin, some of the questions which are raised by this evolving technology go to the heart of the differences between science and technology (and their relative functions in our society), the role of the university (as one major institution of society), and changing relationships both within the university and between the university and the rest of society. Each is examined in turn.

The Changing Relationship between Science and Technology

Weingart (1978) has described the changing roles of science and technology, which have a long and intertwined historical relationship. In the pre-history phase of human development it is possible to trace these activities to very different areas of praxis. Science has its roots in explanation, and was long tied to the role of the priests and magicians of early societies. Technology, on the other hand, is rooted in the practical, everyday experience of the craft trades and artisans.

As we move into later periods, particularly within the evolving university setting, in what might best be described as the pre-paradigmatic period (i.e., before the late 1800's), we see the boundaries between science and technology blurring. Many academics, amateurs,and pre-scientists were heavily engaged in both realms of activity.

However, as the nature of the university shifted to an increased focus on separate disciplines and paradigms (i.e., in the mid to late 19th and early 20th centuries), the two spheres of activity began to diverge markedly. Science returned to its primary concerns with explanation, conceptualization and basic understanding; technology to translating concepts and understanding to practical application. Within the university particularly, technology was viewed largely as a useful outgrowth of science, and was concerned only with application of science to products and practices.

At present, the two spheres seem to be reconverging. In many instances, it is becoming difficult to distinguish scientific activities from technological ones. In a recent work Mulkay (1979) points out that the most important distinction may be the social context of the work, rather than research subjects or methodologies (which are often identical in each sphere). How and to whom information is communicated may be a critical divide. If information flow is primarily horizontal, that is, directed to peers, with few or no constraints upon dissemination, is disseminated first and primarily through scientific meetings or journals, we might say that science is being done. If, on the other hand, most communication is vertical or hierarchical to others in the investigator's organization, if reports have to be cleared prior to dissemination, and if there are other constraints on communication, we might call the activity technology.

In many ways this blurring of functions is symptomatic of a number of more fundamental changes in the nature and role of the university. These changes have implications for the generation of risks and hazards.

The Changing Nature of University Research

Ideally (though admittedly, not always so in practice) the role of the university and professors should be to provide a continuous flow of new knowledge and new leaders. To meet these objectives the university must be a place in which challenge, skepticism and an absence of external controlling influences prevails.

It is undeniable that universities, since their inception, have never been insulated from external influences, and in fact have always served the interests of other societal institutions (e.g., the church, the state, etc.). It is also true that corporate interests have long been influential in university life (e.g., the role of petrochemical industries in U.S. agriculture colleges). However, with changing budgetary realities (i.e., decreases or shifts in priorities in federal and state support for higher education and research), corporate support and influence is becoming more significant.

Moreover some basic values within the university are changing. For example, in the past criteria for academic success included discovery of new knowledge, publication, superior teaching and the like. To an increasing degree (and motivated largely by the financial concerns of university administrators), a new criterion of academic success is the ability to attract research funds and/or commercial offers, and to make money.

These changes imply some very fundamental potential conflicts of interest between the goals of commercial interests and those of the university. The basic goals of commercial interests are to make money and beat the competition. In order to attain these goals private sector firms utilize secrecy and proprietary control of information and research results. By contrast the role of the university (again, ideally) is education and the pursuit of knowledge. The mode should be the open exchange of ideas and the free flow of information and results.

Can these two sets of goals be accommodated within the institution of the university? When private firms invest in universities, they are generally not interested in basic research or general education. Rather they are interested in directed, applied research. This influence alters some fundamental relationships within the university, and can be illustrated by focusing on the area of biotechnology.

First of all (and rather obviously), not all areas of the university are equally positioned to receive funding for such activities. One effect of the increasing prevalence of, and dependence on, these types of funds is to further marginalize the social and behavioral sciences and the humanities within the university hierarchy.

Second, even within the sciences (or the biological sciences, specifically), not all laboratories or professors are equally receptive to such funds. However, as financial resources become increasingly scarce, internal pressures may force faculty to pursue these kinds of moneys. This, in turn has several effects. Students (understandably) will gravitate to those labs with resources (and by necessity to those topics with funding). Some topics will be pursued, and others (which might be equally worthy) will be neglected. Research topics ought (in an ideal sense) to be selected on the basis of potential breakthroughs for new knowledge. However, to a greater degree than ever, hot topics are determined on the basis of commercial marketability.

Third, all of this leads to a new set of roles for university faculty. This new role and its ties to the generation of risks and hazards is explored next.

The Role of the Scientist/Entrepreneur

Just as the distinctions between scientists and technologists are blurring, the role of the scientist or technologist as separate from the entrepreneur is becoming more difficult to discern. This is occurring as activities are pushed by the assumption that accelerated technological advance will produce ever more benefits for society. This blurring is also motivated by the increasing interpenetration of the university by the corporate world (for reasons just described). As Florig states "If knowledge is power, then marketable knowledge is economic power" (Florig, 1986:98).

How is this connected with risk? This situation creates scientists with an economic stake in the outcomes of their research. This, in turn, may lead to an overestimation of benefits, and an underestimation of risks and costs. In the ideal university portrayed above, there is an institutional role for disinterested assessment and appraisal of new technologies and their full range of impacts (including potential risks). If, however, that role is eliminated, or filled by individuals who are also (or primarily) concerned to deploy new technologies as soon as possible, that layer of risk assessment and evaluation is effectively removed.

Summary

It should be clear from this example that risks which might arise from the development and deployment of biotechnology will not occur as the result of extraordinary events or accidents. Rather they will emerge from the everyday relationships between some very fundamental institutions in our society. These include the university, the private sector, and the government. By and large the public has been left out of the debate over the risks and benefits of biotechnology (although, and somewhat ironically, the public has funded a great deal of the basic science on which advances in biotechnology have been built).

In order to keep the public on the outside of this discourse, scientists have appealed to academic freedom, claiming that public scrutiny constitutes an infringement of their rights to unimpeded inquiry. It is interesting to note that similar complaints have not been lodged against the corporate or military partners in much of this research. Similarly, corporations have claimed (with scientists' support) that there is nothing to worry about, and that the public should leave these decisions in the hands of the experts.

Conclusion

I believe that a reorientation of the definition of risk and hazard for hazards research is necessary and appropriate. Hazards researchers should turn their attention to some of the many pervasive, ordinary events which confront human beings on a daily basis. Further, I maintain that such risks are not likely to be reduced significantly through a focus on individual or group response. Therefore, I suggest that hazards researchers begin to examine, in a serious way, the social, contextual factors which give rise to risks of particular types in particular ways in order to come to grips with this issue. It is inevitable that such an examination will force hazards researchers to confront the realities of modern societies in new ways, an outcome that can only promise more relevance and meaning for this work.

References

Ashford, N. 1976. *Crisis in the Workplace: Occupational Disease and Injury*. Cambridge, MA: The MIT Press.

Bogard, W.C. 1988. "Bringing Social Theory to Hazards Research: Conditions and Consequences of the Mitigation of Environmental Hazards." *Sociological Perspectives* 31: 147–168.

Brown, H. 1977. "Resilient Poor vs. Vulnerable Rich." *Ekistics* 44 (260): 4–8.

Cliffe, L. 1974. 'Capitalism or Feudalism? The Famine in Ethiopia.' *Review of African Political Economy* 1: 34–40.

Florig, D. 1986. 'The Scientist Entrepreneur and the Paths of Technological Development.' Chapter 4 in M.L. Goggin, ed., *Governing Science and Technology in a Democracy*. Knoxville, TN: The University of Tennessee Press.

George, L.N. 1982. "Love Canal and the Politics of Corporate Terrorism." *Socialist Review* 12: 9–38.
Hewitt, K., ed. 1983. *Interpretations of Calamity*. Boston: Allan & Unwin. (See especially Chapter 1 'The Idea of Calamity in a Technocratic Age.')
Hunt, V.R. and W.L. Cross. 1975. "Infant Mortality and the Environment of a Lesser Metropolitan County." *Environmental Research* 9: 135–151.
Johnson, B.B. and V.T. Covello. 1987. *The Social and Cultural Construction of Risk: Essays on Risk Selection and Perception*. Dordrecht: D. Reidel Publishing Company.
Kirby, A., ed. 1990. *Nothing to Fear: Risks and Hazards in American Society*. Tucson, AZ: University of Arizona Press.
Mulkay, M.J. 1979. *Science and the Sociology of Knowledge*. Boston: Allan & Unwin.
Nelkin, D., and M. Brown. 1984. *Workers at Risk*. Chicago: University of Chicago Press.
Norwood, C. 1980. *At Highest Risk*. New York: McGraw-Hill Book Company.
O'Keefe, P., K. Westgate, and B. Wisner. 1976. "Taking the Naturalness Out of Natural Disaster." *Nature* 260: 566–567.
Page, J.A., and M.W. O'Brien. 1973. *Bitter Wages*. New York: Grossman Publishers.
Palm, R. 1990. *Natural Hazards: An Integrative Framework for Research and Planning*. Baltimore: The Johns Hopkins University Press.
Perrow, C. 1984. *Normal Accidents: Living with High Risk Technologies*. New York: Basic Books.
Scott, R. 1974. *Muscle and Blood: The Massive, Hidden Agony of Industrial Slaughter in America*. New York: Dutton.
Sewell, W.R.D., and H.B. Foster. 1976. "Environmental Risk: Management Strategies in the Developing World." *Environmental Management* 1: 49–59.
Sjoberg, L., ed. 1987. *Risk and Society: Studies of Risk Generation and Reactions to Risk*. London: Allan & Unwin.
Stellman, J., and S. Daum. 1973. *Work is Dangerous to Your Health*. New York: Pantheon.
Tinker, J. 1984. "Are Natural Disasters Natural?" *Socialist Review* 14: 7–25.
Torry, W.I. 1978a. "Natural Disasters, Social Structure and Change in Traditional Societies." *Journal of Asian and African Studies* 13: 167–183.
______. 1978b. 'Bureaucracy, Community and Natural Disasters.' *Human Organization* 37: 302–308.
Uhl, M., and T. Ensign. 1980. *GI Guinea Pigs*. New York: Playboy Press.
Wallick, F. 1972. *The American Worker: An Endangered Species*. New York: Ballantine Books.
Waterstone, M. 1989. 'Risk Analysis and Management of Natural and Technological Hazards: A Social/Behavioral Science Perspective.' In Y.Y. Haimes and E.Z. Stakhiv, eds., *Risk Analysis and Management of Natural and Man-Made Hazards*. New York: American Society of Civil Engineers.
Waterstone, M., and W.B. Lord. 1989. "How Safe Is Safe?" *National Forum* 69 (1): 22–25.
Watts, M. 1983. "On the Poverty of Theory: Natural Hazards Research in Context." Chapter 13 in K. Hewitt, ed., *Interpretations of Calamity*. Boston: Allan & Unwin.
Weingart, P. 1978. "The Relation Between Science and Technology–A Sociological Explanation." In W. Krohn, et al., eds., *The Dynamics of Science and Technology*. Boston: D. Reidel.
Wisner, B., P. O'Keefe and K. Westgate. 1976. 'Poverty and Disaster.' *New Society* 37: 546–548.

PART I

Risk, Science and Public Policy

This first section presents an assessment of the major issues involved when scientific information interacts with public policy making. The roles of scientists, policy makers and the general public in developing science-based public policy are explored. Several concepts fundamental to an understanding of risk and uncertainty – risk analysis, risk assessment and risk management – are examined, as well as their influence on science and policy making.

In Chapter 1, William Rowe sets out to address the issue of risk analysis from a policy-making point of view. Beginning with some basic definitional matters, Dr. Rowe then offers some examples of policy issues involving risk analysis, and ends with an examination of several approaches which might be taken to risk analysis in a policy context.

In Chapter 2, Helen Ingram, H. Brinton Milward and Wendy Laird explore the ways in which risk issues are situated on policy agendas (the first necessary step toward policy attention). The chapter begins with an overview of the agenda setting process, and identifies the factors which either allow or prevent issues from achieving salience in the political world. The authors utilize the case of global warming to illustrate these steps, and to explore the role of scientists in influencing the agenda setting process for science-based issues.

This first section ends with a practitioner's point of view. In Chapter 3, Dr. Norman Petersen evaluates the utility and problems of risk assessment from the perspective of a state health policy maker.

Chapter 1

Risk Analysis: A Tool for Policy Decisions

W.D. ROWE

Introduction

Increased public and regulatory concern with risks imposed by technological undertakings has focused attention on risk analysis as a tool for aiding in risk-based decisions. Promulgating health, safety and environmental regulations; addressing product and environmental impairment liability in the light of recent adverse court decisions; qualifying new chemicals, pesticides and technological facilities to meet regulations are but a few of the many areas that require new tools and approaches to identify and sort out the many issues involved. Risk analysis, encompassing the assessment and management of risk, has been promoted as one such approach.

In some areas risk analysis has been used quite effectively in both the assessment and management of risk. In these cases, the adequacy of the data base and the decision criteria encompasses the inherent uncertainty in the risk analysis process in a robust manner. However, the situations where adequate data are available and uncertainty ranges are narrow, are rare, resulting in severe limitations to the use of risk analysis in a technical or scientific sense. These limitations to risk analysis are real, and can result in misspent resources and costly mistakes in lives and credibility.

This paper addresses risk analysis from the policy making point of view as a means of identifying the opportunities for and limitations of risk analysis; and how to avoid the limitations. A fundamental precept in the use of risk assessment as a policy tool is: finding the method to solve the problem, rather than finding the problem to fit the method.

Before proceeding, it is important that I state my definition of risk: "*risk* is the down side of a gamble." A *gamble* implies a probability of outcome, and the gamble may be involuntary or voluntary, avoidable or unavoidable, controllable or uncontrollable. The total gamble in which the risk is imbedded must be addressed if the risk is to be analyzed, both the upside (benefits) and down side. Further, I define *risk assessment* to mean the estimation of risk, and *risk management* to mean the reduction or control of risk to an "acceptable" level whether or not the level can be explicitly set. In reality these two processes,

M. Waterstone (ed.), Risk and Society: The Interaction of Science, Technology and Public Policy, 17–31.

which together constitute risk analysis, are not separable since the uncertainty in one affects the judgments we make about the other and vice versa. They may be separated in practice for convenience, but the uncertainties in each area may be the dominant factors in any analysis of risk.

This leads to my definition of risk analysis. *Risk analysis* is a policy analysis tool, using a knowledge base consisting of scientific and science policy information, to aid in resolving decisions. Risk analysis is thus a subset of decision theory, and its importance and utility derive from the uses to which it is put and how well the decisions involved were resolved.

Examples of Some Major Policy Issues Involving Risk Analysis

Three particular examples provide some insight on the scope of policy issues. The first addresses the way risk analyses are carried out, the second addresses some aspects of cancer policy, and the third addresses the separation of risk assessment from risk management.

Use Versus Approach

Two different ways of looking at risk analysis derive from the use to be made of the analysis and the approach to be used to carry out the analysis. These are listed in Table 1.

Table 1. Two Alternative Ways of Addressing Risk Analysis

Use Basis
Spectrum of Uses
Regulatory Analyses
Management Analyses
Public Awareness
Scale of the Analysis
Generic
Specific
Target Audience
Public
Stakeholders
Technocracy
Risk Analysts and Peers
Decision Makers
Approach Basis
Top-down Versus Bottom-up Risk Analysis Approaches
Quantitative Versus Qualitative Health Based Risk Analysis
Probabilistic Versus Consequential Approaches
Absolute Versus Relative Risk Assessments

The Spectrum of Uses of Risk Analysis

There is no such thing as *the* risk analysis. There is a whole spectrum of risk analyses based, at least in part, on the use to be made of the risk analysis. Tables 2a, 2b and 2c identify a spectrum of different uses of risk analysis.

Table 2a. I. Regulatory Analyses

A. Types of Analyses Conducted by Regulatory Agencies

1. *Screening Analyses* – To determine if a risk exists and is high enough to be considered for regulatory control.
2. *Regulatory Impact Analyses* – To justify regulatory actions and satisfy administrative law requirements.
3. *Compliance Analyses* – To demonstrate regulatory violations.
4. *Responding Analyses* – In response to judicial and legislative challenges.

B. Analyses Made by Others in Response to Existing Regulations

1. *Environmental Impact Statements*
2. *Permitting Requirements*
3. *Compliance Monitoring*

C. Analyses Made by Others to Defend Against Unwarranted Regulatory Action

1. *Response to Requests for Comments by Regulators* – Industry response to agency above.
2. *Support of Judicial Actions*
   a. Response to improper agency actions.
   b. Defense against enforcement proceedings.

Table 2b. II. Management Support Analyses

A. Marketing

1. *Absolute Risk* – Demonstrate that a product or process is safe or harmless on an absolute risk basis, that is, the risk on an absolute basis is below some standard or regulation implying an acceptable level of risk.
2. *Relative Risk* – Demonstrate that a product or process is relatively safer and less harmful than alternative and competitive products or processes.

B. Planning

1. *Research and Development*
   a. *Risk Reduction* – Identify areas of high risk (or relatively high risk) in particular products or processes to:
      1. Forestall the need for regulation.
      2. Reduce exposure to future liability claims.
      3. Develop defensive strategies to bound risk liability.
      4. Identify new markets for risk control technology.
   b. *Improve Analysis Capability*
2. *Cost-Effective Use of Resources* – Focus resources on the most risk reduction for a dollar.
3. *Evaluation of Alternatives* – Systems or processes.

Table 2b. (Continued)

C. Risk Management

1. *Prevent Risks from Occurring* – By anticipating and controlling them.
2. *Reducing Exposure* – For health and safety and financial risks for a given, existing process or product. Conduct analyses for:
   a. *System Safety* – Reduction of risk within a system.
   b. *Product Safety and Liability* – Reduce exposure to legal proceedings.
   c. *Third Party Assumption of Risk* –
      1. Insurance – as a means to hedge against risks.
      2. Malpractice – laws to limit liability.

Table 2c. III. Public Education

A. Increase Public Awareness

1. *Seek Rational Public Responses* – A knowledgeable public will hopefully act on information rather than preset beliefs.
2. *Fulfill Regulatory Requirements for Public Disclosure* – A good, simplified and accurate disclosure can also be a useful educational tool.

B. Anxiety Factors

1. *Bring Perceived Risks More Closely into Alignment with Objective Risks* – Anxiety reduction; may also be a defensive strategy.
2. *Frighten People into Action or Agreement* – An offensive strategy attempting to stir fear and anxiety.

This range of uses is neither collectively exhaustive nor mutually exclusive. Table 2a lists some regulatory uses of risk analysis, Table 2b management uses, and Table 2c analyses used to affect public awareness. The point to be made is that each use requires a different type of risk analysis, and the different analyses are not interchangeable.

A second aspect of use is the level of specificity of the analysis. Table 3 provides an illustration of the range of analyses from the specific to the general, using the example of a risk analysis of alternative energy systems. Each level requires a different kind of analysis.

Table 3. Classification of Risk Analyses by Scope of the Application – Micro to Macro Classification

Site Specific Studies
Utility Planning Studies
Power Grid Planning Studies
National Energy Supply Planning
Global Planning
International Energy Planning
Special Purpose Applications
Energy Subsystem Investment
Evaluation of Potential Problems in New Energy Sources
To Support or Reject an Energy Option

A third aspect is the particular target audience or audiences for whom the risk analysis is intended. The public and lay stakeholders require a *presentation* that is different from the technical audience including risk analyst peers, and from the decision makers (who may or may not be technically oriented). If the presentation is different, the analyses themselves for each audience may be different.

Conclusively, the use determines the method, not the other way around. Picking the method to resolve the problem defined by the use is more important than developing universal methods.

Environmental Cancer Risk Policy

Causative Mechanisms for Cancer. Although the cause of cancer is yet unknown, three theories have predominated in recent years (Anderson, 1978: 74). The first is the virus theory. Many or most varieties of cancer, some medical experts contend, are caused by the invasion of a virus whose DNA (deoxyribonucleic acid) system takes over the function of a normal cell, eventually inducing changes in surrounding cells and becoming the start of a cancerous growth. There is evidence from research that certain viruses can induce tumors in experimental animals.

The second theory infers a statistical relationship between cancer and chemicals, poisons, and pollutants in the environment. The marked increase in a specific type of lung cancer in persons who work with asbestos has been well documented. A rare type of liver cancer has begun to be reported among workers who handle certain plastic and chemical substances. The relationship between smoking and cancer has been well established (LeMaistre, 1988). There is no uncertainty that cigarettes and tobacco products are the cause of about 30 percent of all cancers (Loeb et al., 1984). The incidence of cancer on a map of the United States so drawn as to indicate a county-by-county rate of cancer is visibly concentrated in industrialized counties or those associated with high levels of potentially toxic chemicals used in farming and mining. This theory is the basis for the risk analysis of hazardous material disposal practices and toxic chemicals in the environment that has been the basis of cancer risk policy for EPA, OSHA, FDA and other government agencies. It should be noted that there has been no empirical evidence to support the theory that chemicals that cause cancer at high doses in animals will cause cancer in humans at very low doses. We only have unverified theoretical models.

The third theory regards the relationship of stress, tension, and negative emotions to cancer. Some early researchers contend that attitudes of passivity, hopelessness, helplessness, and self-effacement may render persons cancer-prone (Bahnson, 1969). Recent research on this topic has been of two kinds. The first line of research includes experiments with laboratory animals (usually rodents) and has examined the tumor growth effects of stressors such as electric shock, handling, crowding, and restraint. The second has involved studies of the impact of naturally occurring human events (e.g., death of spouse, medical school examinations) on immune function parameters that may be related to cancer (Redd and Jacobson, 1988).

Results from the animal studies are complex and, in certain respects, contradictory. Laboratory research with animals has supported the hypothesis that stress affects tumor promotion, but has no effect on tumor initiation. That is, stress can play a role in whether (and how rapidly) the cancer progresses, but does not appear to influence directly initial cancer cell transformation (Redd and Jacobson, 1988). In animal studies it has been shown that inability to avoid or control stressors (e.g., electric shock) exacerbated tumor growth whereas an ability to escape the shock did not (Sklar and Anisman, 1979). Stress is an adaptive response in which the body prepares, or adjusts, to threatening situations (Sklar and Anisman, 1980).

In humans, stimulation of the hypothalamus and areas of the surrounding brain is associated with reduced psychobiological defense. Changes then occurring through the autonomic nervous system controlled by the hypothalamus, hormonal changes in the adrenal glands controlled by the pituitary and hypothalamus, and biological reactions in the immune system all interact with bio-organisms which can cause malignancy (Bahnson, 1969). It is this diminished immune resistance that can lead to cancer (Fox, 1981). There is evidence that emotions can lead to a variety of different cancers. Blood cancer, lung cancer, breast cancer, ulcerative colitis, cancer of the cervix, and cancer of the prostate have all been related to emotions as a causative factor. These attitudes can be brought on by anxiety.

Although this research is preliminary in many ways, a variety of psychosocial stressors has been found to alter immune function presumed to be related to the body's natural defense against the growth of tumors. These include: death of spouse, medical school examinations, and social isolation. Although these results identify possible mechanisms by which psychosocial factors may effect cancer progression, they are only correlational.

Research is needed to determine if there is an increased incidence of cancer in people who are under chronic stress (e.g., air traffic controllers). At this time a *direct* causal connection between psychosocial stress and cancer in humans has not been demonstrated (Redd and Jacobson, 1988). There are, of course, still other theories about causes of cancer. Genetic susceptibility and diet are but two examples. Moreover, they are not mutually exclusive, all may have some basis as a causative mechanism and may be important to study and understand. The relative dominance of *any* of the causative mechanisms has not been established empirically.

However, in terms of addressing public cancer policy, the comparative importance of the second and third theories above, environmental toxicants versus emotional stress, provides some useful insight into the implications of premature adherence to a single view.

Policy Perspectives on Mis-Information About Cancer Risks. We should be concerned that the mis-information about risks and risk analyses of environmental toxicants might in itself cause higher levels of cancer than those being analyzed. Anxiety in the population affected by a proposed action for which a risk analysis was made and either mis-interpreted or purposefully mis-construed may itself cause an increase in the incidence of cancer. Anxiety is an emotional state of vague uneasiness that is related to either an injury or the threat of an injury as perceived by an individual. After anxiety comes fear and then panic (Anderson, 1978: 74). Anxiety creates stress in the human body and as Dr. Harold Wolff points out, the stress that occurs from any given situation depends on how it is perceived by the individual (Wolff, Wolff and Godell, 1968: 8).

The point to be made is that we are just as uncertain that exposure to the low levels of effluent from a hazardous waste site or other sources of environmental toxicants will cause cancer as we are that increased anxiety by overstating risks will cause cancer in humans. It is not suggested that reasonable critiques and arguments about risks should not take place. Quite the contrary, open, rational discussion about risks is an important part of the public policy decision process. What is considered objectionable and risky is the unfounded and insupportable statements made by those who either have not examined the situation in depth or attempt to influence public environmental policy by unwarranted scare tactics. It is the misuse of information about risks that may itself create more risks than those to be averted.

Separation of Risk Assessment and Risk Management

There is considerable effort expended toward formalizing the separation of risk assessment and risk management into two different spheres. The first that of the scientist and the second that of the policy maker. While it is possible to effect such separation in some cases, generic adherence to such a rule can lead to the masking of critical policy issues. The manner in which uncertainty is addressed in risk assessment is a major policy issue for those who must make decisions using the risk assessment. Conversely the manner in which uncertainties in the evaluation of risks are handled must be of concern to the risk analyst. The very large uncertainties in risk analysis make total separation of the risk assessment and management processes impossible. This does not mean that there are not aspects of risk assessment that should be left to scientists and parts of risk analysis that should be left to decision makers. However, the uncertainties and how they are addressed dominate.

Uncertainties. There are several types of uncertainties involved in risk analysis. These include:

Random Fluctuation	– Inherent random processes
Measurement Uncertainty	– Precision of the measurement system
	– Accuracy of measurement

Interpretation of Measurements	– Objective versus subjective probability approaches – Data inclusion/exclusion judgments – Difference in judgment in meaning of results, e.g., half-full or half-empty – Biases of experts
Model Choice Uncertainty	– Unverifiable uncertainty in the selection of alternative models Biological models Extrapolation models Statistical models
Margins of Safety	– Use of margins of safety to account for uncertainty – Aggregation of margins of safety

The first two types (random fluctuation and measurement uncertainty) are well established, and there are established methods for analyzing these uncertainties. The resulting uncertainty when experts disagree about the interpretation of measurements, is not as well understood; and often is not even recognized as a major source of uncertainty. Moreover, differences in interpretation can be the result of differences in training and methodology used; and, unfortunately, biases if based upon other than scientific motivation.

One aspect of model choice uncertainty is illustrated in Table 4. Seven different types of models are shown that are required to make an estimate of environmental risk, specifically in this case for a toxic organic compound, for radio-iodine and for nitrogen dioxide. The range of uncertainty due to the choice of plausible, available alternative models is shown. The bottom of the range represents situations with low variability, e.g., a diffusion model for a flat even terrain. The higher values are for cases with high variability, e.g., a diffusion model operating on hilly terrain with shear wind strata. When these ranges are aggregated, they do so multiplicatively; and, since they are uniform distributions, their endpoints must be used. As can be seen from the multiplicative ranges in the table, the ranges vary from about two orders of magnitude to fifteen.

With this kind of uncertainty, risk analysis cannot be very robust unless the means of reducing risk are so effective as to overcome these margins of safety. A good example is a hazardous waste facility with a destruction removal efficiency of 99.9999 percent (for PCBs and TCDDs) and a diffusion factor that results in dilution to one part in ten million. This results in a total risk reduction of thirteen orders of magnitude, enough to overcome most of the uncertainty due to model choice. Conversely, a hazardous waste land fill cannot account for more than three orders of magnitude of risk reduction, and cannot overcome the effect of the aggregated margins of safety.

Table 4. Model Uncertainty for Environmental Releases of Three Contaminants to Air

Step	Model	Uncertainty Factor Range		
		Toxic Organic	Radio-Iodine	Nitrogen Dioxide
1. Source Term				
	Averaging in Time	1.1–3	1.1–2	1.1–10
	Averaging in Space	1.1–3	1.1–3	1.1–5
2. Air Pathways				
	Diffusion Models	2–10	1.1–2	1.1–3
3. Metabolic Pathways				
	Organ Intake Models	2–10	1.1–3	1.1–3
	Breakdown Product Models	2–4	1.1–2	1.1–2
	Retention Models	2–4	1.1–1.5	1.1–1.5
4. Dose Estimate				
	Exposure Time Profile	2–10	1.1–10	2–10
	Maximum Exposed vs. Average	2–10	1.1–10	2–10
5. Dose Response Relationship Model				
	Extrapolation – Animal to Man			
	– Scaling Model Choice	40	2–10	20
	– Metabolic Differences	2–100	1.1–1.5	2–5
	Extrapolation – High to Low			
	– Model Choice*	1000	3	100
	– Margins of Safety*	10–1000	2	2–10
6. Individual Risk Estimate**				
	Real vs. Hypothetical Individuals	4–20	4–20	4–20
7. Population (Collective) Risk**				
	Integration vs. Averaging Model	2–10	2–5	2–10
Multiplicative Ranges (Individual Risk)				
	Low (Product of Minimums)	5×10^7	2×10^2	5×10^4
	High (Product of Maximums)	1×10^{15}	4×10^6	3×10^{10}

* Use either, but not both. The first is for non-threshold dose effect relationships, the latter for threshold types.

** Use either, but not both. Either individual or collective risk estimates.

Some Different Approaches to Risk Analysis

One of the fundamental policy issues is the paradigm under which a risk analysis is carried out. Choices of paradigms that overstate the risk because of particular models or conservatisms can lead to expensive programs for reducing risks. Paradigms that understate the risk may lead to actual risks which may be higher than desired. Therefore it is necessary to consider a number of different paradigms from a policy viewpoint. These include top-down and bottom-up approaches, quantitative and qualitative approaches, and probabilistic and consequential analyses.

Top-down and Bottom-up Risk Analysis

The policy analysis aspect of risk suggests that the problem drives the methodology of the analysis. This implies the need for an approach that starts with the problem and works downward to the methods to be used in the detailed analysis. This is called the top-down approach. Table 5 provides definitions of top-down, bottom-up and joint risk analysis approaches.

Table 5. Definitions

Top-down Risk Analysis
The Process Whereby the Risk Analysis Methodology is Tailored to the Policy Needs, and its Feasibility Determined.
Bottom-up Risk Analysis
Taking Each Event That Can Occur in a System and Analyzing the Pathways Leading to the Range of Possible Consequences, and Aggregating These Over the Total Spectrum of Events and Their Associated Probabilities.
Joint Risk Analysis
The Limited Bottom-up Risk Analysis Carried Out as a Result of a Top-down Risk Analysis, and Subsequently Merged With the Top-down Analysis to Form the Final Policy Analysis.

Top-down Risk Analysis. Top-down risk analysis is a paradigm for determining the most appropriate risk analysis for a given decision situation, for making visible the key decision parameters and value judgments involved, for identifying viable alternative strategies for resolution of issues, for identifying scientific and other information critical to the decision process as well as the needed precision of such information, and for communicating the decision process to those affected. The top-down analysis tailors the risk analysis to resolve the issues at hand, and this aspect of the analysis does not necessarily analyze the risks. The top-down approach shows whether it is possible to resolve the policy issues by a subsequent risk analysis; and, if not, identifies the value conflicts that prevent issue resolution by other than political means.

Bottom-up Risk Analysis. In contrast, bottom-up risk analysis starts from the basic scientific information, and attempts to use this information for policy analysis by way of prescriptive methodologies. In nearly all cases, problems arise from large uncertainties in the scientific information base. These problems are usually addressed by retaining and aggregating the ranges of uncertainty, most often in a semi-qualitative manner, or by use of the value judgments of experts or groups of experts. In the first case, the uncertainties are often too large for the analysis to result in meaningful conclusions; or, if forced to conclusions, makes assumptions that mask the uncertainty. The second case substitutes science policy for science. This does not imply that all bottom-up analyses are not useful, but often the resources entailed in making such analyses are very large; and the results are often inconclusive, especially when such an analysis purports to serve all policy purposes.

Joint Risk Analysis. The risk analysis that is made as a result of a top-down risk analysis need only use information necessary to resolve the decision (if it is resolvable by other than political means), and the information used must only be as precise as is necessary for resolution. The combination of a top-down risk assessment with the needed (and only the needed) bottom-up analyses are called the joint top-down, bottom-up approach. It is this paradigm which is preferred, since it joins the best of the two approaches. This approach has been used effectively in a number of different areas, but there is no reason to believe it is any better than other well planned approaches. It is provided to illustrate the different perspectives involved.

Quantitative Versus Qualitative Health Risk Analysis

Quantitative Health Risk Analysis. The National Academy of Science (NAS), Environmental Protection Agency (EPA), Occupational Safety and Health Administration (OSHA) and a number of other federal and state agencies subscribe to a particular paradigm for estimating the risks of exposure to toxic chemicals. I have termed as quantitative, the approach to extrapolating results of bioassays at high dose in laboratory animals to relative potencies at low doses in humans. There is no empirical evidence supporting the determination of relative potencies of two or more chemicals for inducing cancer in humans from laboratory animal data without human epidemiological data. Therefore, the methodology is based upon a number of biological and stochastic models which can only be validated at high dose levels in animals and limited epidemiological studies in human targets of opportunity as a result of historic or accidental exposure. Pharmacokinetics studies of the response of a chemical and its metabolites in the various species under test (including humans when ethical to make such tests), can help in adjusting the biological models, but still do not allow actual validation of the models.

Qualitative Health Risk Analysis. Qualitative approaches attempt to establish no observed effect levels (NOELs) in test animals and provide margins of safety, for example a factor of 100, for acceptable levels of exposure to humans. The inherent model assumptions are that the test animal and humans are metabolically similar, that a no-effect threshold exists, and that an adequate margin of safety below the threshold will account for uncertainties.

In either case the uncertainties should be displayed, not masked. If the total range of uncertainties is displayed, quantitative risk analysis may be a useful means for displaying what is known and what is *not known* about risk. However, it may not be a useful technique for regulating risk in the manner recommended by the NAS, EPA, OSHA, etc. A case by case examination of the problem is probably required rather than a procedural process minimizing the need for exercise of judgment by those using the procedures. Generally, this qualitative process has been used in spite of the lip service given to quantitative risk analysis.

Probabilistic Versus Consequential Risk Analysis

Probabilistic Risk Analysis. Since risk involves a gamble, probabilistic outcomes are inherent. However, the measurement of probability is based upon belief in the meaning of probability. Objectivists base probability estimates on either the frequency of outcomes of previous trials (frequentist approach) or predetermined models (*a priori* approach). Subjectivists believe that they can make judgments about probability, and use these judgments in conjunction with data to improve their estimates (Bayes approach). These fundamentally different approaches are not reconcilable, and these different views spill over into risk analysis.

Probabilistic risk assessment stresses the measurement of probabilities. If the events occur frequently enough, probabilities of future events can be inferred from past experience. Event, fault, and decision trees and statistical decision theory provide useful techniques for estimating the absolute risk of alternative outcomes. "Absolute risk" is defined as the expression of risk levels in terms of probabilities and consequences for a gamble.

However, when events are rare, that is, they occur so infrequently that actual data cannot be obtained, the use of probabilistic analysis to define risk on an absolute basis breaks down. The amount of information that may be inferred from sparse events depends upon belief about probability, specifically whether one is an objectivist or a subjectivist. The frequentists believe that the measured data contain all the information available. The subjectivists believe that additional data can be obtained. The estimates of absolute risk are thus based upon belief structures, and are unverifiable.

There are several approaches used to get around this fundamental problem, including:

System Structures	Measurements on frequent events in subsystems are used to project overall system behavior.
Data Interpolation	Measurements in other systems thought to have identical or similar behavior are used to expand

the data base for inference about the system for which few data are available.

Use of Judgment of Experts — The judgment of analysts is prevalent in probabilistic risk analysis (Apostolakis, 1990). While such expert judgment has always played a role in engineering work, the unavailability of significant statistical or experimental support for most risk situations, makes the use of these subjective estimates attractive.

These approaches have major limitations in application, e.g., common mode failure in system structuring and invalid pooling of data in data interpolation. The use of expert judgment is subjective and is seldom acceptable to objectivists. For example, the use of Delphi techniques to get a consensus of expert opinion in the absence of data is often cited by subjectivists as a means of harnessing expert opinion. The objectivists argue that all one gets is a consensus of opinion that has nothing to do with reality. The most extreme opinion is just as likely to represent reality as the consensus (Abramson, 1981). Moreover, it has been shown that experts are often overconfident in their ability to estimate reality (Fischhoff, Slovic and Lichtenstein, 1977). Additionally it has been demonstrated that the manner in which one frames the problem can affect belief about probability (Tversky and Kahnemen, 1981). The use of judgment of experts itself is a process that has yet to be subjected to scientific verification.

As a result, the application of probabilistic risk assessment for rare events on an absolute basis is not meaningful to most people (perhaps with the exception of some analysts who are extreme subjectivists). Because the assessments depend upon the beliefs about probability held by the analysts, they constitute a doubtful process upon which to base either scientific or public policy.

On the other hand, relative risk analysis of rare events can be useful. The term "relative risk" is defined as the estimation of the risks of alternative gambles in relation to each other. This estimation has less error than absolute risk estimates since many of the errors in the latter estimate are identical for all alternatives, and wash out for the relative case. Probabilistic risk analysis can be used to evaluate the relative risk of alternative systems, to identify vulnerabilities to failure in alternatives or component systems, and to estimate the cost-effectiveness of relative risk reduction among alternatives. Having made a relative risk analysis, a handle on absolute risk may be obtained by "pegging" one alternative to an absolute measure. Pegging involves making an absolute risk assessment for the most well known alternative as a comparison standard. While there will be high uncertainty in the absolute estimate of risk, stakeholders who have experienced a familiar or similar gamble can substitute their own perception of the absolute risk for the "peg". The other alternative may then be scaled upward or downward relative to the peg as it varies. Even though the pegging is "soft" it can be related to the perception of stakeholders as opposed to a belief structure of analysts.

Consequential Risk Analysis. When absolute or relative risk estimates cannot be made effectively for low probability (rare) events, other approaches are in order. Consequential risk analysis assumes that an event occurs, and attempts to prevent a particular outcome from occurring or to mitigate the impact of the outcome. Some of the techniques used involve analytical processes such as sensitivity analysis, optimization theory and cost-effectiveness of risk reduction methods; and contingency processes that mitigate consequences such as emergency response planning, early warning systems, and evacuation operations.

Conclusion

There is no such thing as *the* risk analysis for a problem. There is a spectrum of analyses based primarily on the use to which the analysis will be put. The analysis must fit the problem not the other way around. This requires a decision framework for structuring the approach to the problem to be addressed. The most useful decision framework is associated with policy analysis applications and perspectives.

Structuring the problem (using the variety of different, available approaches) before selecting methodologies for solution is a requirement for making useful risk analyses from a policy perspective. The top-down approach illustrated here is one such process, particularly when addressing policy decisions. Pursuit of method independent of application assumes that the methods will have wide application. This is seldom true. More often, these methods have limited application for a specific purpose or area, and therefore, one must identify the application and find the methods most useful. Use what works, discard what does not.

The dominant factor in determining whether an approach will fail is the total uncertainty in the risk analysis. If the uncertainties in the risk assessment are greater than those in the acceptance criteria used to make decisions, then the risk assessment is useless, despite the fact that analysts may claim otherwise.

Uncertainties are the limitation in risk analysis. They dominate all aspects of the analysis in both the assessment and management aspects. The opportunities come in finding the right approach for the particular problem, and the particular manner in which the approach is used to solve the problem for a specific policy decision.

References

Abramson, L.R. 1981. "Some Misconceptions About the Foundations of Risk Analysis." *Risk Analysis* 1(4): 229-230.

Anderson, R.A. 1978. *Stress Power.* New York. Human Sciences Press. New York, 1978, p. 74.

Apostolakis, G. 1990. "The Concept of Probability in Safety Assessments of Technological Systems." *Science* 7 Dec. 1990, p. 1363.

Bahnson, C.B. 1969. "Psychophysiological Complementarity in Malignancies." *Annals of the New York Academy of Sciences* 164(2): 319–333.

Fischhoff, B., P. Slovic, and S. Lichtenstein. 1977. "Knowing With Certainty: The Appropriateness of Extreme Confidence." *Journal of Experimental Psychology: Human Perception and Performance* 3: 552–564.

Fox, B.H. 1981. "Psychosocial Factors and the Immune System in Human Cancer." In R. Ader, ed., *Psychoneuroimmunology*. New York: Academic Press, pp. 103–158.

LeMaistre, C.A. 1988. "Reflections on Disease Prevention." *Cancer* 62(8), October 15, 1988 Supplement.

Loeb L. et al. "Smoking and Lung Cancer: An Overview." Cancer Research.

Redd, W.H., and P.B. Jacobson. 1988. "Emotions and Cancer: New Perspectives on an Old Question." *Cancer* 62: 1871–1879.

Sklar, L.S., and H. Anisman. 1980. "Social Stress Influences Tumor Growth." *Psychosomatic Medicine* 42: 347–365.

Sklar, L.S., and H. Anisman. 1979. "Stress and Coping Factors Influences Tumor Growth." *Science* 205: 513–515.

Tversky, A., and D. Kahnemen. 1981. "The Framing of Decisions and the Psychology of Choice." *Science* 211: 453–458.

Wolff, H.G., and G. Wolff, and H. Goodell. 1968. *Stress and Disease*, Springfield, IL: Charles C. Thomas.

Chapter 2

Scientists and Agenda Setting: Advocacy and Global Warming

HELEN INGRAM, H. BRINTON MILWARD, AND WENDY LAIRD

Introduction

The important contribution that scientists make to agenda setting has not been sufficiently recognized in scholarly literature. The students of agenda setting have only recently come to realize the importance of ideas that come from sources outside the government. There has been little research that identifies the scientific community as a source of agenda topics. While there is a large literature devoted to the relationship between science and government, it has generally been concluded that there is no scientific elite with an undue influence upon government. Instead, the principal concern has been the perversion of science by politics. The activity of subgroups of scientists acting as entrepreneurs to set the public agenda on topics in which they have a strong interest has not been widely acknowledged.

The role of science entrepreneurs in introducing, popularizing, and elevating the idea of global climate change to national agenda status illustrates that, contrary to much of the literature, science can play a critical role in agenda setting. The central section of this paper recounts the history of how global climate change became a "hot" political topic and the contribution made by science activists. To place the case study in context, the paper first reviews two relevant literatures: the centrality of experts and ideas in agenda setting; and the assessments of scientists' influence in policy. The chapter concludes with a revision of prevailing notions of the constraints upon science activism and an outline of the conditions under which scientists are likely to become important in agenda setting.

Experts and Ideas in Agenda Setting

The literature on agenda setting while diverse falls into two broad patterns (Nelson, 1984: 22–25). One pattern focuses on organizational approaches and stage theories of agenda setting. Works displaying this pattern focus on the actions of decisionmakers and the organizations they command in relation to

M. Waterstone (ed.), Risk and Society: The Interaction of Science, Technology and Public Policy, 33–53.

the stages of the agenda setting process (Cobb and Elder, 1972; 1983; Cobb et al., 1976; Nelson, 1984; Walker, 1977). The second pattern is the issue cycles, careers, and clusters pattern. Anthony Downs (1972) and more recently Peters and Hogwood (1985) attempted to determine if issue cycles consisted of predictable stages as both the public and government attend to them. For our purposes the issue careers and clusters aspects of this second pattern are more germane to our research on global warming than issue cycles. We wish to focus on "... the mobilization of groups and individuals around an issue" (Nelson, 1984: p. 25).

Central to this second pattern of agenda setting research is the work of Donald Schon (1971). The importance of his work lies in his treatment of ideas as innovations and the role that functional networks play in their diffusion. He believes that ideas obey a law of limited numbers; there are many more ideas than can ever be dealt with on the public agenda (Schon, 1971: 123). Schon's interest lies in which ideas are chosen and why. Does a type of coupling occur (an idea with an interest) to pull the idea into good currency and onto the public agenda? As these ideas surface, networks of individuals and interest groups gravitate to and galvanize around the new ideas helping to frame both the issue and its solution (Van de Ven, 1986: 592). These actors provide resources and energy that move the idea onto the public agenda. Later scholars would call this a policy community (Walker, 1981; Kingdon, 1984; Nelson, 1984; Laumann and Knoke, 1987; Cook and Skogan, 1989).

Much of the early stages of the agenda-setting process occurs in policy communities made up of policy professionals who work in government, think tanks, universities, research centers, and interest groups. Individuals within the policy community know one another and interact on a regular and consistent basis regarding the issues within the purview of their policy domain. Policy domains include such areas as health, the environment and energy. In a policy dense world, these policy communities play a key role in agenda setting. They consist of that set of interested parties that revolve around issues in an area of professional, or policy relevance. The policy community can be tightly knit, as in the case of health, or loosely linked, as in the case of energy (Laumann and Knoke, 1987).

While policy subsystems have been a standard part of political science since the 1950s (Maass, 1951; Redford, 1960; and Freeman, 1955) rarely has this type of analysis been applied to scientific communities. The role that scientists play in setting the agenda, particularly their role as policy entrepreneur, or the role the policy communities of science play in creating agenda items for public debate, has not been sufficiently recognized. While scientists have always had knowledge, it is only recently that power has been attributed to them. Jack Walker was one of the first to explore this relationship (Walker, 1981). He looked not at scientists but professions. He attempted to link increases in the knowledge base in the profession with the ability of the policy community to create issues for the public agenda.

Critical to the process of creating an issue is the role of events, policy entrepreneurs and the increase in scientific and technical knowledge in an area. Events cannot be predicted with precision and often produce impacts different than those first anticipated. Events can serve as a "policy window" (Kingdon, 1984) that if the entrepreneur has prepared the ground well, will thrust the issue onto the public agenda and lead to policy action. An event like Chernobyl, may in a day, raise the issue of the safety of nuclear power to the forefront of the public agenda. Similarly an issue may emerge as the consensus judgment of the community, such as the problem of child abuse which arose from the community of medicine. Policy entrepreneurship takes much longer than events to raise issues to the agenda. However, when an issue does emerge there are organizations and individuals who have become supporters of the issue and make it more likely that the issue will be addressed. Policy entrepreneurs behave like entrepreneurs in the private sector. They have a vision and seek to implement it against great odds, and in the process, overcome entrenched interests and hostile competitors.

Increases in scientific and technical knowledge create tensions in a policy community between the conventional wisdom of the field and the new knowledge or technology that calls it into question. Within a given scientific community, it is possible to observe policy entrepreneurs using findings or models opposed to the conventional wisdom to convince their colleagues of the validity of their issue by publishing results in scientific journals, being interviewed by science writers from major papers, magazines, or television networks, and testifying before congressional committees.

To summarize, the agenda setting literature has come to recognize the importance of policy communities, policy entrepreneurs, and ideas in agenda setting. Studies have only recently begun to examine the contribution of scientists to this process.

Assessments of Science Influence

Scientists often portray themselves as outside the political process and as poorly understood by politicians. They complain that the objectively determined best solutions to problems as revealed by science seldom carry much political weight. Scientists maintain that even when they receive a hearing, their views are distorted by the press and misinterpreted by political actors. The scholarly literature in political science and public policy presents a contrary view. Researchers find that scientists wield a great deal more influence than they admit to having, especially in environmental policy. Yet, most scholarly assessments of contemporary science's activity in public policy also find there are clear constraints to the scientists' exercise of influence.

Growing Policy Involvement

Since at least the end of World War II, scholars have recognized that natural scientists played a significant role in public policy. Science made a critical contribution to the war effort, and the incredible success achieved throughout the war by university scientists turned weaponeers, convinced many military officials to continue a close liaison in the post war years. Certain politicians viewed scientific research as a means of promoting full employment after the war. Further, scientists were successful in convincing many public officials of the importance of basic research as a precursor to the miracles for modern life that were ultimately to come from science (Gilpin and Wright, 1964: 6).

By the mid-1960s, political scientists were heralding scientists as the new apolitical elite. The exponential rise in public research and development funds; the establishment of new governmental agencies and programs with heavily scientific and technological missions; and the rapid evolution of mechanisms such as science advisory committees and expansion of the National Academy of Sciences, provided close links between scientists and government. In an edited book published in 1964, Don K. Price, author of *The Scientific Establishment,* wrote:

> Yet the plain fact is that science has become the major establishment in the American political system: the only set of institutions for which tax funds are appropriated almost on faith, and under concordats which protect the autonomy, if not the cloistered calm, of the laboratory (Price, 1964: 20).

Robert C. Wood identified the enormous respect which the American culture has for science and technology as a key asset that heightened science advisors' influence in comparison to other skill groups such as lawyers. In his view the ambivalence and ignorance of the public concerning scientific affairs was less important than the widely-held conviction that science offered a "way out." This public perception was bolstered by scientists who portrayed their enterprise as an endless frontier, the riches of which were only beginning to unfold for society (Bush, 1945).

Political scientists recognized that the efficacious self conception of science provided a powerful impetus for political activity. Wood observed that spokespersons for science had articulated at least three conceptions of the scientific enterprise designed to establish its social value and public usefulness. First, science is described as innocent inquiry, the results of which can be used for good or evil as determined by politicians rather than the discoverer. Second, science is crucial to the three great public goals: national security, conquest of disease, and increase in public welfare. Finally, scientists portrayed themselves as specially endowed to bring order and sense out of the political process. Reluctantly, but dutifully they emerged from the laboratory and sacrificed their professional careers in the interests of an informed debate on great public issues. Effective policy making required just what scientists believed they had to offer: objective sifting of the facts, balanced visions, thoughtful reflection, and the mobilization of the best wisdom and highest competence (Wood, 1964: 64).

In the mid-1960s, some political scientists were arguing that scientists had power to actually create public issues. Don Price described the policy making process as "a process of interaction among the scientists, professional leaders, administrators, and politicians; ultimate authority rests with the politicians, but the initiative is quite likely to rest with others including the scientists in and out of government" (Price, 1965: 967–68).

The motivations for science activism were identified with self interest, but extended to broader governance concerns. According to Dorothy Nelkin, scientists focussed efforts on creating a positive public image with the press when demand for research funds began to far outstrip the supply. Today academic institutions, research organizations, government agencies, and corporations involved in costly scientific and technological developments have all increased their public relations activities in order to enhance institutional prestige, encourage public support for research, and influence public policy (Nelkin, 1987).

Even individual scientists have tried to attract public attention to influence public policy, to attract funds, or to establish their competitive position in a "hot topic." Nelkin notes that DNA researchers initiated a remarkable media campaign to show that genetic engineering research was safe, that its critics were irresponsible, and that regulation was unnecessary (Nelkin, 1987: 138).

Some scientists are clearly motivated by ideology and a concern for public affairs. One British computer scientist explained his speaking out against the Strategic Defense Initiative as a consequence of the failure of government to adequately reflect scientific understanding. He believed that even though the technical impossibility of the computer aspects of SDI were clear to computer scientists, this knowledge would not be sufficient to stop this project. The scientist stated, "The threat this represents to all our futures is so great that in the end one must, in the old Quaker phrase, 'speak truth to power' " (Thompson, 1985).

Scientists and Environmental Issues

Students of environmental policy have long recognized the contribution of science and individual scientists to the evolution of the issues. Outside the area of national defense, scientists have had the greatest influence in environmental policy. Both the definition of problems and the formulation of solutions to problems of environmental quality, including pollution and environmentally imposed health risks, have depended upon science and technology. For instance, the very definition of safe drinking water has hinged upon a breakthrough in technology to measure minute traces of elements, and advances in health sciences to relate exposure to risk.

Scientific entrepreneurs have had a definitive role in shaping the environmental agenda and influencing policy action. The historian Donald Fleming gives great credit to Barry Commoner for legislation which mandated civilian control of atomic energy after World War II and exposing the potential danger of atomic testing. Commoner moved on from atmospheric testing to embrace ecology, the environment, and pollution. Commoner worked in tandem with the media to keep issues such as the "killing of Lake Erie" before the public and public officials. He also targeted detergents as the source of phosphates causing eutrophication of lakes and streams, although later findings confirmed that providing better water treatment, not modifying soap was the most efficient course of action (Fleming, 1972).

While Commoner has been one of the quintessential science activists in environmental policy there have been others, including Rachel Carson, Paul Ehrlich, Carl Sagan, and Garrett Hardin. Such figures have purposely targeted an audience much wider than their peers in popularizing the imminent dangers of such threats as the population bomb, the silent spring, and the nuclear winter. Their foci have been the lay person and the policymaker. They have also reached considerably beyond their data to suggest appropriate courses for public action.

Concerns Related to Science Activism

Interestingly, while scientists and many contemporary scholars have observed the close relation between science and government, the concern has been to protect science from dangerous perversion, not to control scientists' undue influence. Scholars have argued that there is little chance that scientists constitute some sort of privileged elite who can impose their own interest on policy for a number of reasons.

Science activism is constrained by the rules of good science. Most scientists believe that their status with other scientists is a good deal more important than their public image. Scientists will jealously guard their independence, objectivity, and accuracy. Maintaining a scholarly reputation involves thorough peer review of findings before they are publicized. Research is seen as a collective effort in which scientists carefully build upon knowledge previously accumulated. The rules of good science work against the oversimplification of complex issues that accompany the politics of public policy. Even those who do not faithfully follow the rules recognize them. According to Stephen Schneider:

> Scientists toil for years in some specialty trying to uncover some of nature's mysteries. They publish their findings. Over time those publications contribute to the most precious intangible a scientist ever owns – his or her scientific reputation. The unwritten rules in science decree that recognition is supposed to be based on years of careful work backed up by scores of publications appearing in the most strictly peer-reviewed scientific journals dealing with narrowly defined topics. Published deeds that stand the tests of time are supposed to build one's recognition, not clever phrases that capture the public's – or worse – the media's – attention (Schneider, 1989: 200–201).

The establishment of science often sharply disciplines individuals that stray too far from established rules. For instance, the science activists in environmental policy do not come from mainstream science, and have undoubtedly made career sacrifices. Fleming notes Commoner was never a biologist of the first rank. His (partly ideologically motivated) arguments with molecular biologists, over the properties of DNA badly damaged his professional reputation (Fleming, 1972). Further, repeated involvement in politics harms the scientist's reputation for objectivity. The more an activist becomes identified with one end of the environmental policy spectrum, the more he or she is likely to be referred to as an environmentalist rather than a scientist.

The post World War II fears of a priesthood of science coming to dominate politics has all but disappeared due to the changes in the structure of science. At the same time as science was gaining power after World War II, its institutional base was becoming more fragmented. University based academic science has been supplemented by science in industry, interest groups, and government. The diversification of the institutional basis of interests among scientists has its correlate in the loss of a common frame of value-orientations and beliefs as well as a common basis of interests among scientific and technical experts (Weingart, 1982). Further, the competition for diminished research funding operates to referee excessive scientific claims. The individual scientist who reports unsubstantiated findings is called to account by competitive scientists, and the field that exaggerates its importance invites rebuttals from scientists in other fields.

From the perspective of scientists, their influence in public policy is sharply limited by media misunderstanding and distortion. In 1980, for example, Philip Handler, then president of the National Academy of Sciences wrote, with reference to environmental disputes, that the news media were perniciously infiltrated with anti-science attitudes (Nelkin, 1987). The penchant journalists have for controversy can cause them to try to find a spokesperson for the other side, even when there is no other side. Scientists do not like to think that the quality of their science is likely to be challenged in the pages of newspapers rather than through peer review.

Rather than concerning themselves with the political power of scientists, many scholars have written about the vulnerability of science to politics and government, and have worried that scientific agenda may become politically driven. As scientific research has become incredibly expensive and substantially funded by government, scientists are encouraged to involve themselves in examining questions that government will find useful. Because the state holds the purse strings for the research establishment in the United States, it can fairly easily command the advice of science for policy making. Scientists may have very little control over how advice is used. One observer of science in relation to environmental policy wrote:

> Environmental issues place scientists in a highly charged political atmosphere where impartiality and objectivity, among the most highly esteemed scientific virtues, are severely tested and sometimes fail. Scientists are often consulted by public officials in good part because the scientists' presumed

> objectivity, as well as technical expertise, makes them trustworthy advisers. But objectivity may be an early casualty in the highly partisan and polarizing atmosphere of policy conflict. Even if a scientist can maintain impartiality, he/she can not prevent partisans of one or another policy from distorting technical information to their own advantage (Rosenbaum, 1985: 95).

In a 1989 book, the sociologist Chandra Mukerji portrayed science as used by government. She argues that in exchange for support and relative autonomy in intellectual pursuits, scientists have given up their voice. They are kept on tap by government to give advice that will avoid grave technical and scientific errors, and to identify scientific opportunities. Their testimony is also used to bolster policy alternatives that are mainly chosen for reasons unrelated to science. She writes:

> The process of giving the voice of science to the state for its political ends is in formal terms the opposite of ventriloquism. Scientists do not send their voices out to speak through the mouths of mute government officials. Government officials extort the language of science and scientists' analytic skills to do their political jobs. Scientists are made mute except for when politicians find their voices useful (1989: 198).

That science may not speak with a single voice about matters is not all that damaging to the usefulness of advice to government. The value of science to government is not primarily in the truth of what scientists say, but in the legitimacy they provide. If an idea is backed strongly by a reputable scientist, it can not be easily dismissed. The fact that other scientists may disagree is to be expected since everyone knows that it is possible to find respected researchers with opposing views on many topics. Politics and the public have little understanding or respect for the process through which an idea is tested against prevailing scientific skepticism before it is thought by scientists to be scientifically established.

To summarize the scholarly assessment of the role of science in public affairs, there is a widespread acknowledgment of the growing interrelationship, but very little concern about excessive science influence on any stage of the policy making process. If anything, scientists are seen to be more vulnerable to being used by government than the other way around.

The Agenda Status of Global Climate Change

As a topic on the public agenda, the "greenhouse" issue arrived in June, 1988. The issue burst into public consciousness in a manner reminiscent of the first Earth Day which mobilized environmental awareness in 1970. The "triggering" event was James Hansen's testimony before the Senate Committee on Energy and Natural Resources. Hansen, a scientist at the National Aeronautics and Space Administration testified in June that the warming trend witnessed during two hot, dry summers was indicative of a larger warming trend. Hansen testified that there was a 99 percent chance that global warming, ascribed to increased levels of carbon dioxide (CO_2) had already occurred. His testimony raised the ire of many in the scientific community who believed his position was premature. Consensus had not been reached, and many scientists felt the traditional, lengthy peer review process had been subverted. This scientific dispute and the tampering of Hansen 's testimony by the Office of Management and Budget, account for the flurry of media coverage, and the movement of the issue onto the public agenda.

In analyzing the public agenda status of the greenhouse issue, three indexes were used: the National Newspaper Index, the Popular Magazine Index and the Vanderbilt T.V. News and Archive. We believe these databases represent popular or mass public viewer and readership and hence, are indicative of public agenda status. As an example, the number of articles devoted to a subject in the National Newspaper Index, which indexes such papers as the New York Times and the Los Angeles Times can be used as a yardstick for public or systemic agenda status. Figure 1 displays the annual number of articles devoted to the greenhouse issue. (Database constraints prohibited searches earlier than 1979.) Through 1987, the number of articles remains under twenty. By 1988, however, there is a nearly tenfold increase (McKibben, 1989).

Accompanying this media exposure came a virtual flood of Congressional activity. In the second session of the 100th Congress, thirty two bills were introduced, including the Global Warming Act (S 608), the Stratospheric Protection Act (S 491), and the World Environmental Policy Act (S 201). Twenty eight days of hearings were held by nine committees (Ingram and Mintzer, 1990).

The source of the global climate change idea is clearly science; the theory of global warming and the attempt to empirically validate it go back at least 100 years. Svante Arrhenius, in the late 1890s employed measurements of infrared radiation to calculate the possible effects of carbon dioxide on the earth 's temperatures. He concluded that average global temperatures would rise as much as nine degrees fahrenheit if the amount of CO_2 in the air doubled from pre-industrial levels. This "hot house" theory had been explored nearly a century before by Jean-Baptiste Joseph Fourier who speculated that carbon dioxide, a by-product of fossil-fuel combustion, trapped solar infrared radiation, and similar to a greenhouse, heated the earth 's lower atmosphere.

Table 1 outlines the slow, incremental process of accumulated scientific knowledge that undergirds the global climate change issue since Arrhenius published his findings. Fossil fuels were connected to increasing CO_2 levels in the atmosphere in 1925 (Lotka, 1925). In the 1930 's, a link was found between industrial production of CO_2 and temperature (Callendar, 1938). In 1957, the ocean was no longer assumed to provide a large enough sink for CO_2 associated with fossil fuel burning (Revelle and Suess, 1957). By 1958, scientists were sufficiently concerned with the CO_2 problem to establish two monitoring stations, Mauna Loa, Hawaii and one at the South Pole to monitor the airborne fraction of CO_2 (Keeling, 1987).

By the mid-1970s, CO_2 was no longer deemed the sole culprit in global

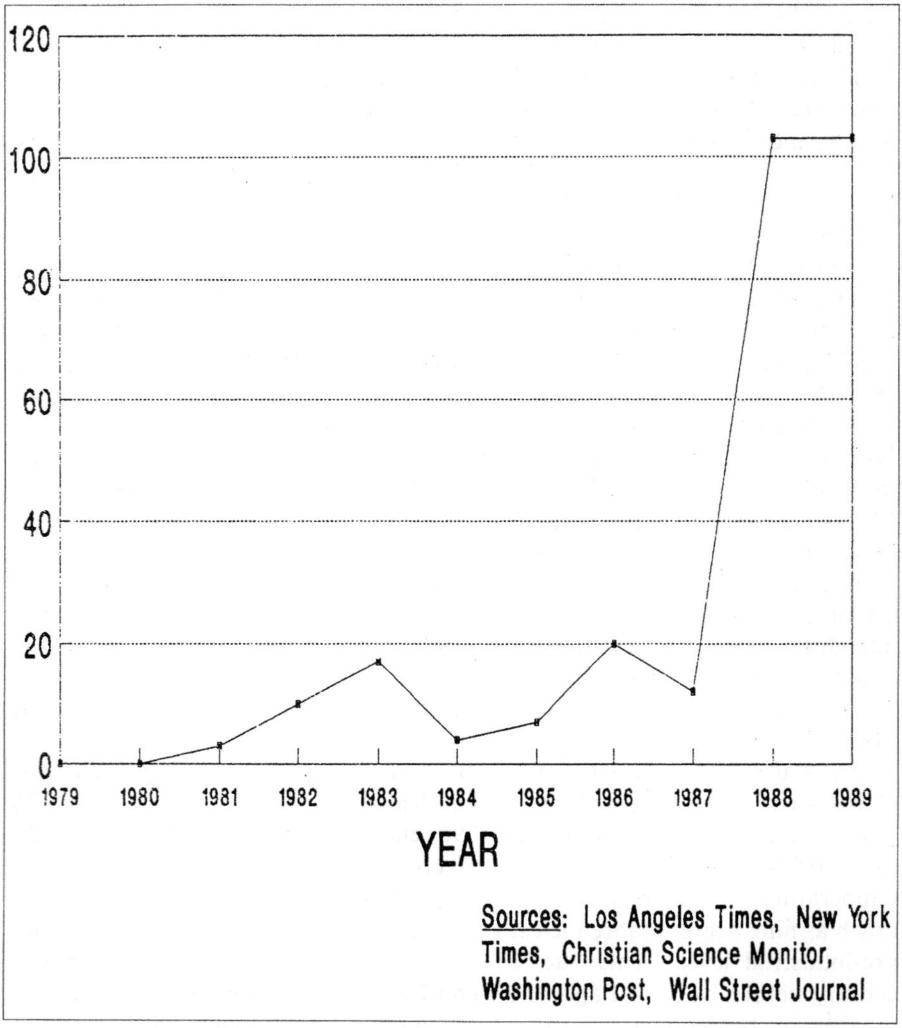

Figure 1. Number of published articles on Greenhouse Effect in five major newspapers (1979–1989).

warming. Other gases such as methane, tropospheric ozone and chlorofluorocarbons (CFC's) were also found to contribute to global climate change (Molina and Rowland, 1974). In 1985, Ramanathan et al., argued that these trace gases were just as important in causing global climate change as CO_2 (Ramanathan et al., 1985). Ice core data, an alarming hole in the ozone over Antarctica, and increased precision in CO_2 measurements represent a slow movement by the scientific community to understand the global climate change problem.

Table 1. Major Scientific Events

1896	S. Arrhenius	First numerical calculations relating CO_2 concentrations with Earth's surface temperature.
1925	A.J. Lotka	Associated industrial use of fossil fuels with increasing CO_2 in atmosphere.
1938	G.S. Callendar	Made direct connection between industrial production of CO_2 and temperature.
1957	R. Revelle, H.E. Suess	Said most CO_2 from burning fossil fuels would remain in the atmosphere and not be absorbed by the oceans.
1958	C.D. Keeling	Established the first CO_2 monitoring station at the Mauna Loa Observatory, Hawaii.
late 1960's		Advent of computer modeling of the greenhouse effect.
1974	M.J. Molina, F.S. Rowland	Associated increased use of chlorofluorocarbons with ozone destruction.
1985	V. Ramanathan et al.	Concluded that trace gases were just as important as CO_2 in determining climate change.
1985	D. Raynaud, J. Barnola	Used ice core analysis to relate increase in atmospheric CO_2 to burning of fossil fuels.
1985	J.C. Farman et al.	Discovered hole in the ozone layer over Antarctica.

Figure 2 displays the relationship between publications on the issue of global climate change in scholarly and disciplinary journals and the media's attention to the issue. Three databases were used to describe the movement of the greenhouse issue to the public agenda, and one database, the Pollution Abstract Index, describes the specialized or scientific literature. Noteworthy, is that the three media databases simultaneously peak in 1988. This simultaneous peak is accounted for in the agenda setting literature, which argues that the popular press, or that produced for mass consumption, reacts rather than leads the scientific literature (Nelson, 1984). In this instance, it would appear that popular publications were being driven by specialized publications. The careful peer review of technical papers (pre-publication) and the later translation into the simplified language of the popular press apparently took place in traditional sequence. Quite possibly, scientists writing during 1987–88 were testifying not just to their peers, but to gain a wider, public audience. An alternative explanation is that both scientific and popular press were driven by the accumulated evidence garnered through the years. This relationship was aided by the establishment of science as a regular newspaper

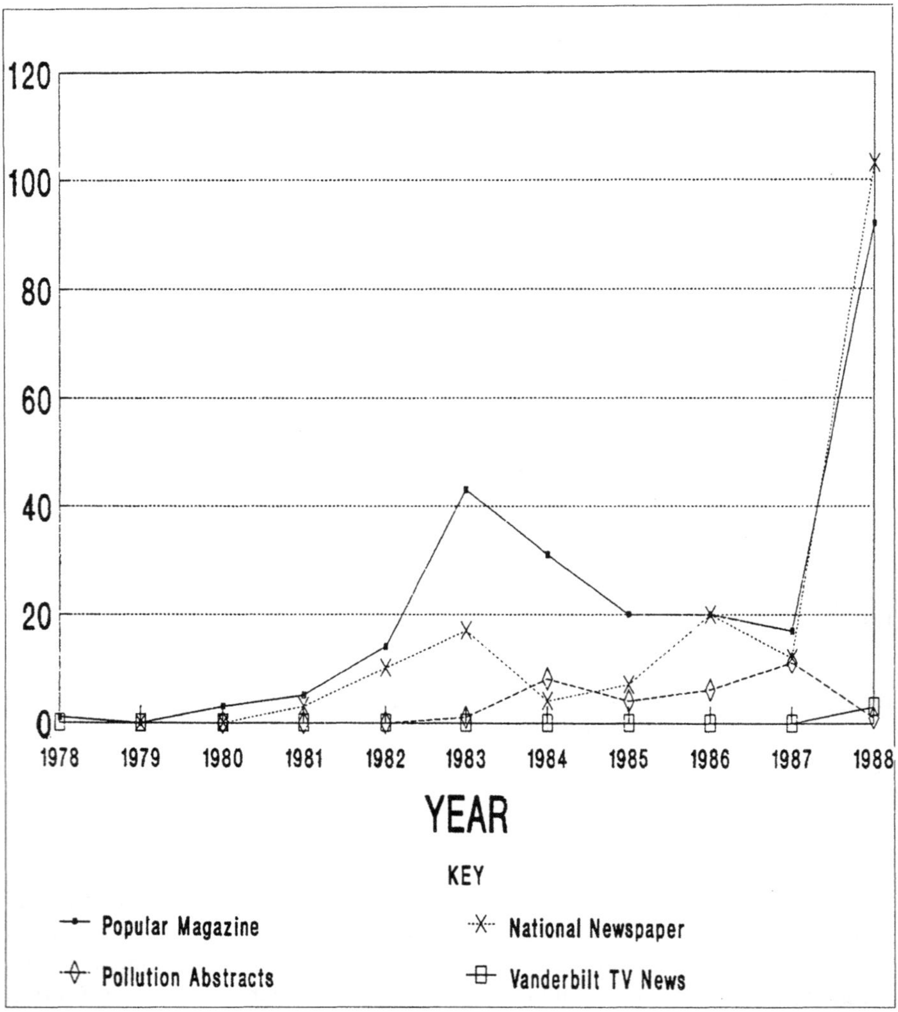

Figure 2. Number of Greenhouse Effect citations (1978–1988).

"beat" in 1970, which helped to form a direct connection between the media and scientists' findings.

Agenda placement cannot be explained by some absolute scientific consensus about the seriousness and certainty of findings. Many of the most recent findings regarding global climate change are confusing and confounding. Much of the measurable increase in urban temperatures, for example, can be traced to the "urban heat island effect." The accuracy of the instruments used to collect data upon which some studies depend also is in dispute. In addition, the amount of CO_2 uptake by the biota and the ocean is poorly understood, and is more complex than current models can predict. Research findings suggest there are many natural variations to temperature and cyclical warming and cooling trends which are unrelated to human activity. Further, scientists have quite different views about the biological and climatic processes that will occur in the face of higher CO_2 levels and/or warmer temperatures. Feedback mechanisms are one of the most confounding of the greenhouse issues. In addition, the models upon which future predictions depend are not powerful enough to emulate today's temperatures. Different models produce different results.

Summing up the implications of what science does and does not know about global climate change in a recent article, two MIT scientists concluded:

> What seems to differentiate scientific interest at the turn of the century from that today is the level of hysteria shared by the press, environmentalists, and, sad to say, many scientists that the whole ecosystem will be unraveling within the next few years. We appear to be in a period when those who cry "wolf" the loudest are those who receive financial backing for continued research. Science, in the word of one observer, has become a "contact sport." ...
>
> There are good arguments for continuing large-scale support for climate change research but no compelling arguments to pursue action programs at domestic or international levels unless those programs can be justified on other grounds, such as the ecological damages caused by massive deforestation. It may be as long as 20 years before we will have definitive answers... (Rogers and Fiering, 1989).

In summary, the slow accumulation of scientific data alone does not explain the bursting of the global change issue onto the public agenda. Neither does the concept of a new "breakthrough" in scientific knowledge, nor does consensus among the scientific community explain its sudden appearance. The spurring of the greenhouse issue onto the public agenda, as a major risk issue, therefore, must be explained by other factors. Few argue that the movement of the greenhouse issue onto the public agenda was due primarily to members within the scientific community acting as policy advocates, deterministically striving to make the global climate change issue an item on the public agenda.

Science Advocates and Science Networks

The fragmentation of science into specialized groups, many members of which operate in research organizations in government and independent organizations outside universities has resulted in sub-communities of relatively small networks of researchers. Those belonging to the global warming research network have special stakes in the placement of climate change upon the political agenda. The competition for research dollars is intense, and the relative support for different subcommunities in science varies with time. Big increases in research support go to scientists researching "hot" topics. Reasons these topics become hot include the likelihood of dramatic breakthroughs in new basic science knowledge; employment of such exciting technology as high temperature superconductors or supercomputers; the close relationship of findings to issues affecting health and welfare, like AIDS or cancer; and the breadth of disciplines involved both inside and outside government.

Global climate change has many of the ingredients to become a "hot" research topic. Large scale global modelling involves "high" science with elegant mathematics and computer operations. Predicting the consequences of global warming offers work to a large number of scientists in many disciplines. Finally, while global climate change is not predicted to come soon, the consequences are portrayed as very grave: coastal flooding and widespread drought, loss of wetlands, deforestation, intrusion of saltwater into freshwater rivers, loss of prime agricultural lands, increased skin cancer, and many others.

Advocacy Scientists

Entrepreneurship and advocacy are required to create a hot research topic. A corps of scientists in several key positions were crucial in fashioning a greenhouse interpretation of events such as the heat and drought of 1988. The network of advocacy scientists can be roughly divided into two types according to the nature of their jobs: research center scientists and environmental interest group scientists.

The archetype of research center advocates in global climate change is Roger Revelle. The significant discovery he made in 1957 with H.E. Suess (Table 1) initiated a commitment to gaining attention to global warming that continues to the present. Ominously, in 1957 Revelle wrote, "human beings are now carrying out a large scale geophysical experiment of a kind that could not have happened in the past nor be repeated in the future" (Revelle and Suess, 1957).

Revelle was active in 1980 in putting together a controversial proposal under the aegis of the American Association for the Advancement of Science (AAAS) for an interdisciplinary group to work through scenarios for the implications of climate change. Calculating that climate change might provoke a large public reaction if it were associated with precious water supplies and future scarcity, flooding and drought, Revelle was a key facilitator in putting together the 1987-88 study of global climate change and water resources.

A much younger research center science advocate, Stephen Schneider, is central to global warming's agenda setting activity. From the early years of his career, Schneider deliberately tried to be provocative about climate change. He has pursued not only the discovery of scientific evidence, but has developed a skill in dealing with the mass media and political processes to bring public and policy attention to the issue. While Schneider is a leading climatologist at the National Center for Atmospheric Research, he has also written a Sierra Club book to warn the average citizen of the "unprecedented threat to the global environment" (Schneider, 1989). His book contains an entire chapter entitled

"Mediarology" in which he describes the challenges for a scientist determined to gain media attention for his/her issue. In addition to his book, Schneider is an old hand at Congressional testimony, and is regularly on a lecture circuit that includes foreign capitols as well as college campuses.

While a postdoctoral fellow at the Goddard Institute for Space Studies, Schneider became closely acquainted with James Hansen, the most outspoken of the research science advocates. According to Schneider, The Goddard Institute constantly has been struggling for money. Schneider writes:

> ...by the mid-eighties Jim [Hansen] had testified a number of times before Congress and several juicy quotes from him had made it into the national media. He began to develop the reputation as an ardent believer in the seriousness of the greenhouse effect, and used calculations from his institute team to back up his strong words (Schneider, 1989: 195).

On June 23, 1988, during the heat wave and drought, Hansen made a statement that shocked his colleagues and remains controversial. Speaking of the observed 0.4 degree centigrade warming of the earth's temperatures, which well could have been taken as a chance event, Hansen said, "with 99% confidence, we can state that the warming during this period is a real warming trend" (U.S. Congress, 1988: 39).

Realizing that environmental regulation must be supported by scientific reason, numbers of interest groups have begun to support research staffs during the last decade. Most such scientists have doctorates and are socialized as scientists to be careful and prudent, yet they work for a cause and are hired to give credibility to their group's position. Research center science advocates on the global climate change issue often have institutional homes outside colleges and universities. The degree of dependence of such scientists upon grant funding to support research is important. Scientists in interest groups are another important segment of the global climate change network and have even stronger incentives for advocacy. In the global climate issue, these scientists have come to occupy an active role in preparing reports and giving testimony.

George Woodwell has drawn attention to the rapid destruction of forests that he says will result in the disruption of the global carbon cycle. Among other positions, Woodwell is founder and vice chairman of the board of the Natural Resources Defense Council, as well as founding and current trustee of the World Wildlife Fund.

Michael Oppenheimer, an atmospheric physicist who is senior scientist at the Environmental Defense Fund served on the steering committee which prepared the widely cited Bellagio Report, "Developing Policies for Responding to Global Climate Change." Oppenheimer has appeared with Ted Koppel on "Nightline" as well as before Congress. In 1988, he told a Senate committee, "In my personal opinion, greenhouse warming presents the most important global challenge of the next few decades on a par with defense, disarmament, and economic issues" (U.S. Congress, 1988: 80).

Scientists at the World Resources Institute have been special advocates of the global warming issue. Jessica Tuchman Matthews, a biochemist and biophysicist, has written numerous op-ed pieces promoting increased gasoline taxes as an appropriate response to increased levels of greenhouse gases in the atmosphere. Irving Mintzer, a Ph.D. from Berkeley, who specializes in energy policy is another prominent global warming scientist with the World Resources Institute. His research adopts the framework of policy analysis, and examines the consequences of possible scenarios resulting from different policy alternatives. However, his framework is distinctly different from that which would be sponsored by most academic policy analysts in that problem definitions tend to be taken as givens rather than the starting point for analysis. The options examined are those that are attractive to environmentalists for reasons of environmental quality, apart from global change, rather than an analysis of different approaches that might include human adaptation as well as no action.

Science Writers

Science writers have had a large role in making the greenhouse effect a big issue. Science has become a regular beat in many newspapers and magazines. Eighteen daily newspapers, with a total circulation of seven million, added special weekly science sections between 1977 and 1987. This creates a "news hole" that must be filled on a regular basis. The New York Times, for instance, has twelve science journalists (Nelkin, 1987). There are numerous science writers at national meetings such as those of the American Association for the Advancement of Science (AAAS) searching for novel stories. Stephen Schneider writes about his experience as a speaker at the AAAS meetings on the subject of human activities which could change climate:

> After speaking for half an hour or so on how various kinds of human activities could change climate, I concluded that unfortunately, only a relatively few people were even aware of the possibilities. In the [moment I] quipped: 'Nowadays everybody is doing something about the weather and no one is talking about it'.
>
> At the front of the audience a distinguished looking gentleman was taking notes. He turned out to be the dean of all science writers, Walter Sullivan of the New York Times. Since journalists love one liners – especially if they boil down complicated issues into a quick phrase, or reflect or create controversy, the next day's New York Times featured a story on weather control that closed with my reverse Mark Twain Quip (Schneider, 1989: 200).

The climate change issue has many of the ingredients that make for good copy for science writers. It is related to something that is important and with which everyone has direct experience, the weather. Climate is no respecter of persons or their status and affects the rich, poor, liberals, conservatives, business and labor alike. Portraying the serious potential risks of alterations in the ozone layer, climate, and precipitation is easy.

Further, the climate change issue has a healthy quota of villains. Brazil and other countries with large tropical rainforests can be blamed for destruction of the CO_2 absorbing trees. Coal burning factories and power companies who have long been blamed for acid rain can now be accused of changing climate. Motorists who are polluting urban air can add heating up the world to their crimes against nature.

As is it has developed, global climate change is attractive to science writers because it offers controversy and conflict. Statements by science advocates increasingly have been countered by fellow scientists, some of whom are engaging in a good deal of advocacy themselves. It is now a fairly simple matter for journalists to write a story with two conflicting sides.

Abetting Factors

It is doubtful whether science advocates and their collaborators among science writers would have succeeded in getting global climate change onto the agenda were it not for the summer of 1988. Prior to 1988, the climate change story got to the press primarily when there was a "weather peg," such as a hurricane or drought. Cracked earth, withering plants, stranded barges, record high temperatures, and the supposed link to global climate change dominated the news for weeks in 1988, as is illustrated in Figure 1.

Elected politicians also functioned as issue abettors, although it is clear that the initiative for the new agenda item came from outside government. One of the ways in which elected politicians attract attention beyond their own constituency is to associate themselves with issues that have the potential of appealing to a wider audience. It is therefore not surprising that Senator Albert Gore, who at age 39 had already served in the House and Senate and declared himself a Presidential candidate, was among the earliest legislators to see the potential in the global climate change issue. While a member of the House of Representatives in 1981, Gore came to the rescue of AAAS 's interdisciplinary proposal for research on the impact of global climate change. When the monitors for the Department of Energy project proposed to eliminate funding, Gore suggested that increased efforts should go forward to reduce the areas of scientific uncertainty. Even at that early date, Gore perceived that the evidence for global warming was a matter of near agreement among the mainstream of atmospheric scientists. Gore displayed none of the tolerance for caution in interpretation and differences in view that characterize the scientific peer review process. In fact, he has accused researchers who present findings at odds with global warming, of being irrelevant (Michaels, 1989).

Furthermore, Gore has not wanted to admit that there might be any "winners" in global warming. He has said, "Those who have been involved in the debate for a long while have come to see the winners and losers phrase as an impediment to increased awareness and a bar to meaningful action" (Schneider, 1989: 258).

Politicians have not been alone in attempting to "couple" their interests and careers with the climate change issue. Besides environmental groups, many government agencies and some business groups have allied themselves to lessening the causes, mitigating the consequences, or making accommodations to the inevitable. Agencies have a stake in framing the issue to maximize the opportunity of increasing their share of the budget, their chances of survival, to press their policy objectives, or bolster their power and prestige within the bureaucracy. Energy and chemical companies are concerned that the issue may foster new regulations. They also hope that protecting the ozone layer and conserving climate may somehow prove lucrative. Just as pollution control is a business serving a number of large and small enterprises, climate change may well attract its own clientele (Ingram, Cortner and Landy, 1990: 433–438).

Overview of Agenda Setting in Global Climate Change

The source of the idea of global warming is clearly in the scientific community, and the responsibility for placing it upon the public and governmental agenda rests primarily with a network of scientists who had a clear interest in drawing attention to the potential risks. The global change advocates based in research centers and interest groups had an easy target in selling their idea to science writers. The idea of global climate change has many of the attributes that make for good copy. The issue could be portrayed as having serious and risky consequences that affect peoples' lives.

It is not possible to ascribe the sudden emergence of the global climate change issue to the march of science. There were no dramatic breakthroughs in evidence occurring as the issue was popularized in 1988; and in fact, the issue was becoming more confusing and controversial in scientific terms.

Implications for Literatures on Agenda Setting and Science and Public Policy

Only a few of the large number of possible topics that could be addressed by the public and decision makers actually make it onto the legislative agenda for action. The number of topics that can be taken up at one time is clearly limited [i.e., Schon's notion that ideas obey a law of limited numbers (Schon, 1971: 123)]. Each topic that finds its way to the agenda has crowded its way above other possible items. Therefore, the power to originate ideas and to attract public and political leaders' attention to them is significant.

Agenda scholars have come to recognize the importance of ideas and the mobilization of issue networks through the attraction of ideas in agenda setting. Further, scholars have noted that ideas often come from the periphery, that is outside government and interest group officialdom. However, the extent to which science networks and science entrepreneurship can be the source of agenda itself is not sufficiently appreciated. While it is impossible to generalize from the single case of global climate change, the example does point to the significant resources scientists bring to agenda setting. In a political environment where many political institutions have lost public confidence, citizens continue to have great faith in science and technology to interpret the world objectively. Further, many of the areas of government activity have a strong technical or technological component, particularly those related to environmental protection and public health risks. Citizens and public officials depend upon scientists to signal when air is clean, when water is safe, and currently, when the climate is warming.

The scholars who have studied the relationship between science and public policy during the past twenty years have not generally recognized the power of science over agendas. While writers have been concerned about the extent to which government funding of science research might distort research priorities, there has been too little notice of the extent to which such funding provides incentives for scientists to lobby government on behalf of their own subject areas. The rules of good science have not worked to significantly restrain such science advocacy. The science establishment is no longer sufficiently cohesive for its discipline to be completely effective. Scientists now find themselves in many different kinds of institutions, including interest groups and centers heavily dependent on grants, which pursue goals potentially conflicting with conservative and cautious science. There are many pressures to publicize before publishing and to substitute propaganda and argument for peer review.

There has been concern that the close advisory role that scientists have come to play in government has meant that scientists too often lend their reputation for objectivity to legitimize positions and policies determined by government for reasons other than science. It may also be the case that the scientist's reputation for objectivity may be used to legitimize the special interests of some subfields or some scientific networks. In their zeal to obtain what they view as their fair share of attention, science advocates may exaggerate the significance of their findings. In the long run, such exaggeration may damage the overall status of the scientific community as well as distort the overall balance of effort directed to the many important science questions needing answers.

The structure of media has changed to abet the influence of science. Science writers have developed a clear bias toward finding newsworthy stories from scientific discovery. They compete with each other to reveal the most recent findings. Science writers do not share scientists' commitments to the peer review process. Instead, they often interview scientists with opposing points of view in hopes of enlivening stories with controversy.

It may be argued that science contains self correcting mechanisms. Advocates that step beyond the rules of good science are bound to attract public criticism from fellow scientists. It is true that the most outspoken global warming advocates have drawn authoritative rebuttals. There are substantial rewards offered to science advocates who enjoy public notoriety, however, and some global warming network members appear to enjoy controversy. Further, there are overarching risks and costs to public policy. Many other important environmental and other issues may have been crowded off the agenda as decision makers have focused on global climate change. And, the public may have become thoroughly disenchanted and skeptical of all scientific claims about global climate change long before the evidence becomes clear about what warming may be in store and policy action is clearly appropriate. The legitimacy, in the eyes of the public, of the greenhouse issue may be lost as the role of the scientist as policy advocate emerges.

References

Bush, V. 1945. *Science: The Endless Frontier*. Washington: National Science Foundation.

Callendar, G.S. 1938. "The Artificial Production of Carbon Dioxide and its Influence on Temperature." *Quarterly Journal of the Royal Meterological Society* 64: 223.

Cobb, R.W., and C.W. Elder. 1972. *Participation in American Politics: The Dynamics of Agenda-Building*. Boston: Allyn and Bacon.

Cobb, R.W., and C.W. Elder. 1983. *Participation in American Politics: The Dynamics of Agenda-Building* (2nd ed.). Baltimore: Johns Hopkins University Press.

Cobb, R.W., J. Keith-Ross, and M.H. Ross. 1976. "Agenda-Building as a Comparative Political Process." *American Political Science Review* 70: (1): 126–138.

Cook, F.L., and W.G. Skogan. 1989. "Agenda-Setting: Convergent and Divergent Voice Models of the Rise and Fall of Policy Issues." Northwestern University, Unpublished Paper.

Downs, A. 1972. "Up and Down with Ecology: The Issue Attention Cycle." *The Public Interest* 28: 38–50.

Fleming, D. 1972. "Roots of the New Conservation Movement". Vol. 6 of *Perspectives in American History*. D. Fleming and B. Bailyn, eds. Cambridge: Charles Warren Center for Studies in American History.

Freeman, J.L. 1955. *The Political Process*. New York: Random House.

Gilpin, R., and C. Wright, eds. 1964. *Scientists and National Policy-Making*. New York: Columbia University Press.

Ingram, H., H. Cortner, and M. Landy. 1990. "The Political Agenda." In Paul E. Waggoner, eds., *Climate Change and U.S. Water Resources*. John Wiley and Sons, Inc.

Ingram, H., and C. Mintzer. 1990. *How Atmospheric Research Changed the Political Climate*. Tucson, AZ. Udall Center for Studies in Public Policy Issue Paper No. 1.

Keeling, C.D. 1987. *Measurements of the Concentration of Atmospheric Carbon Dioxide at Mauna Loa Observatory, Hawaii*, 1958–1986. Final Report for the Carbon Dioxide Information and Analysis Center. Oak Ridge, TN: Martin-Marietta Energy Systems Inc.

Kingdon, J.W. 1984. *Agenda, Alternatives, and Public Policies*. Boston: Little Brown.

Laumann, E.O., and David Knoke. 1987. *The Organizational State: Social Choice in National Policy Domains*. Madison: University of Wisconsin.

Lotka, A.J. '1925' 1956. *Elements of Physical Biology*. 'Baltimore: Williams & Wilkins.' Reprint. New York: Dover.

Maass, A. 1951. *Muddy Waters: The Army Engineers and the Nation's Rivers*. Cambridge: Harvard.

McKibben, B. 1989. "The End of Nature," *The New Yorker*, p. 58. A review of the National Newspaper Index bears out the fact that 1988 articles focus on the Hansen testimony, OMB's altering of it and the debate.

Michaels, O. 1989. The Greenhouse Climate of Fear. The Washington Post. January 8.

Molina, M.J., and F.S. Rowland. 1974. "Stratospheric Sink for Chlorofluoromethanes: Chlorine Atom Catalyzed Destruction of Ozone." *Nature* 249: 810–812.

Mukerji, C. 1989. *A Fragile Power: Scientists and the State*. Princeton: Princeton University Press.

Nelkin, D. 1987. *Selling Science: How the Press Covers Science and Technology*. New York: Freeman and Company.

Nelson, B.J. 1984. *Making an Issue of Child Abuse: Political Agenda Setting for Social Problems*. Chicago: University of Chicago Press.

Peters, G.B., and B.W. Hogwood. 1985. "In Search of the Issue Attention Cycle." *The Journal of Politics* 47: 238–253.

Price, D.K. 1964. "The Scientific Establishment." In *Scientists and National Policy-Making*. New York: Columbia University Press.

Price, D.K. 1965. *The Scientific Estate*. Cambridge: Belknap Press of Harvard University Press.

Ramanathan, V., R.J. Cicerone, H.B. Singh, and J.T. Kiehl. 1985. "Trace Gas Trends and Their Potential Role in Climate Change." *Journal of Geophysical Research* 90: 5547–5566.

Raynaud, R., and J. Barnola. 1985. "An Antarctic Ice Core Reveals Atmospheric CO_2 Variations Over the Past Few Centuries." *Nature* 315: 309–311.

Redford, E.S. 1960. "A Case Analysis of Congressional Activity: Civil Aviation: 1957–58. *Journal of Politics* 22(2): 228–258.

Revelle, R., and H.E. Suess. 1957. "Carbon Dioxide Exchange Between Atmosphere and Ocean and the Question of an Increase of Atmospheric CO_2 During the Past Decades." *Tellus* 9: 18.

Rogers, P., and M. Fiering. 1989. "Climate Change: Do We Know Enough to Act?" *Forum for Applied Research and Public Policy*, (Winter): 15–13.

Rosenbaum, W. 1985. *Environmental Politics and Policy*. Washington: Congressional Quarterly Press.

Schneider, S. 1989. *Global Warming: Are We Entering the Greenhouse Century?* San Francisco: Sierra Club Books.

Schon, D.A. 1971. *Beyond the Stable State*. New York: Norton.

Thompson, H. 1985. "Why Scientists are Speaking Out." *New Scientist* (21 Nov): 28–29.

U.S. Congress. Senate. Committee on Energy and Natural Resources. Hearing on Greenhouse Effect and Global Climate Change. 100th Cong., 2nd sess. June 23, 1988. S. Hearing 100–461, pt 2.

Van de Ven, A.H. 1986. "Central Problems in the Management of Innovation." *Management Science* 32(5): 590–607.

Walker, J.L. 1977. "Setting the Agenda in the U.S. Senate: A Theory of Problem Selection." *British Journal of Political Science* 7(4): 423–445.

Walker, J.L. 1981. "The Diffusion of Knowledge, Policy Communities and Agenda Setting: The Relationship Between Knowledge and Power." In John L. Tropman, et. al., eds., *New Strategic Perspectives on Social Policy*. London: Pergamon.

Weingart, P. 1982. "The Scientific Power Elite – a Chimera: the Deinstitutionalization and Politicization of Science." In N. Elias, H. Martin, and R. Whitley, eds., *Scientific Establishment and Hierarchies*. Dordrecht, Holland: D. Reidel Publishing Company.

Wood, R. 1964. "Scientists and Politics: The Rise of an Apolitical Elite." In R. Gilpin and C. Wright, eds., *Scientists and National Policy-Making*. New York: Columbia University Press.

Chapter 3

Risk Assessment and the Communication of Risk in a State Health Agency

NORMAN J. PETERSEN

State health agencies constitute one segment of government in which policy decisions are sometimes based on the results of risk analysis. Protection of the public health from the insults of environmental contaminants is a high priority activity that is often accentuated by a high level of public concern. While risk assessment is particularly useful in dealing with issues that involve the exposure of humans to toxic substances and the adverse health effects that may be associated with these exposures, the real challenge is to respond with policies that are sensitive to the public concern as well as capable of resolving risk problems.

The Arizona Department of Health Services, Division of Disease Prevention routinely utilizes the principles of risk assessment in response to two types of problems. The first type involves exposure or possible exposure of a human population to a toxic substance in the environment. Typically, the major concern is to determine whether adverse health effects are likely to occur as a result of the exposure. Obviously, the answer to the question is of great personal importance to the individuals within the population. Additionally, the result of a risk assessment can be used by a health agency to determine what, if any, public health measures are needed and whether further epidemiologic study is indicated.

The second problem area in which risk assessment principles are useful is in the setting of standards or guidelines for toxic substances in such media as water, air, and soil. Health-based guidelines are essential for the effective monitoring of the environment to protect the public health and for determining the extent of clean-up necessary at sites where contamination is known to have occurred.

In 1983, the National Academy of Sciences (NAS) published the results of a study of the institutional means of risk assessment at the federal level (National Academy of Sciences, 1983). In defining the nature of risk assessment it was agreed that the process contained some or all of the following four components:

1. *Hazard Identification*–This component involves gathering and evaluating data on the types of health injury or disease that may be produced by a

M. Waterstone (ed.), Risk and Society: The Interaction of Science, Technology and Public Policy, 55–58.

chemical and on the conditions of exposure under which injury or disease is produced. It may also involve characterization of the behavior of a chemical within the body and the interactions it undergoes with organs, cells, or even parts of cells. The ultimate goal is to determine whether a particular chemical is or is not causally linked to particular health effects.

2. *Dose-Response Evaluation*–This component involves describing the quantitative relationship between the amount of exposure to a substance and the extent of toxic injury or disease. Data are derived from animal studies, or less frequently, from studies in exposed human populations. There may be many different dose-response relationships for a substance if it produces different toxic effects under different conditions of exposure. The risks of a substance can not be ascertained with any degree of confidence unless dose-response relations are quantified, even if the substance is known to be toxic.

3. *Human Exposure Evaluations*–This component involves describing the nature and size of the population exposed to a substance and the magnitude and duration of their exposure. Human exposures to substances in the environment may occur because of their presence in air, water, or food. Other circumstances may provide the opportunity for exposure, such as definite contact with the substance or contact with contaminated soil. The evaluation must distinguish between the amount of the substance in the medium in which it is present and the amount actually absorbed by the subject. The evaluation could concern past or current exposures, or exposures anticipated in the future.

4. *Risk Characterization*–This component involves the integration of the data and analysis of the first three components to determine the likelihood that humans will experience any of the various forms of toxicity associated with a substance. In cases where exposure data are not available, hypothetical risk can be characterized by the integration of hazard identification and dose-response evaluation data alone. This characterization also includes the estimated overall effect of the uncertainties inherent in the first three components.

While risk assessment is a useful tool for public health agencies, the application of public policies developed with this tool are often problematic. At the root of these problems is the difficulty in effectively communicating risk concepts to the public. This difficulty arises because the expectations and perceptions of the public concerning risk can be quite different from those of the health officials attempting to implement a risk-related policy or to explain the health risk associated with exposure to some environmentally mediated toxic substance. Two examples that help illustrate this problem are presented here.

The first deals with the problem of uncertainties that pervade the first three components of risk assessment. Because so few human data are available on the cause-effect relationships of toxic substances and adverse health effects, experimental animal data make up the bulk of information used in risk assessment. Extrapolation of findings in animal studies to humans, especially in the sensitive area of cancer causation, is a subject that remains controversial even among acknowledged experts. Beyond these uncertainties are those normally associated with scientific measurements, statistical models and margins of safety. The health expert trained in this specialty, is able to understand and use the available data fully aware of these limitations.

However, the general public has not had the benefit of specialized training in such fields as toxicology, environmental epidemiology and risk assessment. Further, the public has been conditioned by several generations of public health successes in the control of infectious diseases that were, to a large extent, the result of relative certainty about cause-effect relationships and the effectiveness of preventive measures. Accordingly, the public has a greater expectation of certainty concerning environmentally provoked health risks than science can deliver with the present level of knowledge. This situation is epitomized by the fact that a truthful answer of, "we don't know and may never know," by a health official about some environmental health problem is often unacceptable to the general public.

A second example of the difficulty in communicating risk concepts involves the public's perception of risk. Peter Sandman, a risk communication expert, has pointed out that if one makes a list of environmental risks in order of how many people they kill each year, then list them again in order of how alarming they are to the general public, the two lists will be very different (Sandman, 1987). The reason for this is a difference in the definition of risk. To a public health official, risk means expected rate of disease incidence or mortality. The public, however, views risk as much more than a statistical rate. Experts in risk communication have identified at least 20 additional factors that influence the public's perception of risk. Among these are:

- Natural vs. manmade – natural (acts of God) risks are more acceptable than artificial risks.
- Voluntariness – voluntary risk is much more acceptable than coerced risk.
- Control – a risk over which one has control is more acceptable than risk that is controlled by others (e.g., government).
- Fairness – enduring greater risks than one's neighbors without access to greater benefits produces outrage.
- Familiarity – familiar risks are more acceptable than exotic or high-tech risks.
- Dread – risks from some diseases are more acceptable than the risk from others such as cancer and AIDS.

Because these "outrage" factors are an intrinsic part of the public's perception of risk, they must be considered as important in the communication of risk. An attempt to change this public definition is less likely to be successful than recognizing the fact that the public responds more to outrage than to hazard. A productive dialogue between health officials and the public on the subject of risk will be possible only when both parties agree to use the same definition of risk.

In the absence of a superior alternative, risk assessment will continue to be an important public heath tool in dealing with toxic substances in the environment. The challenge will be to improve risk communication. This will only happen if health officials treat the public with fairness and honesty and with respect for the public 's right to make their own decisions. Through mutual effort and understanding it should be possible to narrow the expectation and perception gaps that often exist between the public and public health agencies on the subject of environmental health hazards.

References

National Academy of Sciences. 1983. *Risk Assessment in the Federal Government: Managing the Process*. Washington, D.C.: National Academy Press. Committee on the Institutional Means for Assessment of Risks to Public Health.

Sandman, P.M. 1987. "Risk Communication: Facing Public Outrage." *EPA Journal* 13(9): 21–22.

PART II

Allocating Scarce Medical Resources

Advancing medical technology offers great promise for extending life, reducing pain and suffering, and eradicating or limiting disease. Transplant and other types of surgery, biotechnologies, life sustaining procedures, and newly developed vaccines and medicines have provided undeniable benefits for those who can obtain them.

However, it is often the case that new techniques and procedures are extremely expensive, and therefore available to relatively few people. It is also true that in many cases the research necessary to develop these new approaches comes at the expense of advances in more mundane kinds of medical care (for example, public health medicine, prenatal counseling, preventive medicine, and others). The fact that medical resources (researchers, funding, laboratory facilities, etc.) are relatively scarce means that trade-offs must be made to allocate these resources efficiently and equitably.

Making such trade-offs raises very difficult legal, economic, social, and ethical questions. For example, how do we, as a society, allocate such resources between rich and poor, between young and old, or between heroic measures for the few and more mundane measures for the many? This section addresses these issues, and attempts to provide some insight into how such trade-offs are made, and what particular trade-offs imply in terms of the health and welfare of citizens.

Albert Jonsen begins this section with an overview of "Modern Medicine as a Risk to Society." In Chapter 4, Professor Jonsen discusses heroic medical interventions (in this case the artificial heart program), and contrasts these technological advances with more mundane kinds of medical needs (e.g., long-term care). Jonsen ends with an examination of budgetary considerations and an exploration of some of the equity concerns in health care distribution.

In Chapter 5, Deborah Mathieu follows with an examination of the particular risks and hazards inherent in the current configuration of the American health care system. Professor Mathieu documents a number of difficulties which prevent access to care, and suggests several approaches that might be used to improve access. She goes on to discuss several sources of discrimination in health care provision, and discusses the particular issues of under- and over-treatment which pervade the present system.

Finally, in Chapter 6, Theresa Cullen, M.D. offers a contemporary, on-the-line perspective of these issues with her discussion of the everyday choices faced by physicians in highly resource-constrained environments. Working with the Indian Health Service, Dr. Cullen has faced the hard issues of health care rationing on a daily basis. Her viewpoint adds a sense of the real-world impacts of these decisions.

Chapter 4

Modern Medicine as a Risk to Society

ALBERT R. JONSEN

The general topic of risk and society may sound a bit strange when applied to medicine and health care. For the last few decades, and indeed, the last century, medicine has generally been seen as the source of great benefits for society. From the middle of the last century, major public health efforts were made to clean up polluted water supplies to reduce the conditions for the serious epidemics of cholera that had ravaged the country, immunization was introduced which has practically eliminated the once devastating smallpox, typhoid fever and polio, and improvements in personal health came in a dramatic fashion with the introduction of antibiotics in the years just immediately preceding World War II. All of these brought enormous benefits to society and prestige to medicine and science. To consider medicine as itself a source of danger is almost counterintuitive. In addition, since the very beginning of Western medicine in ancient Greece, the ethic of medicine has been enunciated in the maxim, "be of benefit and do no harm." Those words, found several times in the medical literature of ancient Greece, have echoed through the centuries as the principle par excellence that should govern the ethical practice of healing (Jones, 1923).

Clearly, however, that ancient maxim bears witness to the fact that healers have always known that their skills could also cause great harm. The invasion of the human body with knives and probes, the amputation of limbs, the instilling of toxic substances are all causes of great harm to a human being. But the ethic of medicine required that whenever such harm was done, it was done with a view to a great benefit, the saving of a life, the prevention of more serious disability. Even when physicians in the past used certain medicines which were themselves quite powerful poisons, those medicines were used with the belief and intent of bringing benefit to those who received them. It has also been an absolute requirement of ethical medicine to repudiate what we call quackery. Obviously in the past it was often difficult to differentiate quackery from orthodox medicine. Nevertheless there has always been throughout the history of medicine, a repudiation of any form of medical activity which was done primarily to exploit a patient, to take advantage of the vulnerability of persons suffering from illness and disease. When we reflect on that long history, is it not

M. Waterstone (ed.), Risk and Society: The Interaction of Science, Technology and Public Policy, 63–73.

peculiar to assert that medicine and medical science is a risk to society and creates risks?

I first realized the possibility of medicine creating risks in 1972 when, having just joined the faculty of the medical school of the University of California, San Francisco, I received a telephone call from Dr. Theodore Cooper, who was at the time the director of the National Heart, Lung and Blood Institute of the National Institutes of Health. Dr. Cooper invited me to join a committee which was being asked to review the ethical, legal, social and economic implications of the totally implantable artificial heart. I had not heard of the totally implantable artificial heart when he offered me that invitation, but he persuaded me that this was an important task because it was the first time that the National Institutes of Health had devised a committee with the special charge of evaluating the ethical implications of a new technology. I learned that the Congress of the United States had established and funded a program for research to develop a totally implantable artificial heart in 1962. That program had been funded at ten million dollars per year and its objective was to bring into being a totally implantable artificial heart by 1970. Dr. Cooper's phone call came in 1972, and the artificial heart was still far from being ready for service, but clearly there was a very vigorous and quite exciting program of development at the National Institutes of Health (Jonsen, 1973).

The totally implantable artificial heart was a mechanical pump and a power source that was to be entirely implanted within the chest of a human being after their natural heart had been totally excised and removed. It was to be a completely internal device, not tethered to an outside power source or pump. The blood pump, which would replace the natural heart, was to be made of plastic and a Dacron-like material. That material would contract 70 to 90 times a minute just as the natural heart does. It would pump approximately five quarts of blood through the body at a constant circulation and relatively constant pressure, just as the natural heart does when healthy. The power source, which for the natural heart is a small bundle of fibers, the natural pacemaker, was to be for this heart, 250 grams of plutonium. Those 250 grams would be encapsulated in a lead capsule about the length of an ordinary pencil and about twice as thick. That capsule would be placed within the abdomen. This lead capsule was thought to be of sufficient strength and thickness to shield the plutonium from causing any excessive radiation. Thus the device was called the totally implantable artificial heart. That is, every feature of that device, the pump and the power source, would be completely internal to the patient.

It became fairly obvious, as our committee began to learn about and discuss this technology that this device was a risk to society. Two hundred and fifty grams of plutonium 238, even when well shielded in a capsule, still has radiation effects. It was quite obvious that the persons who carried the capsule in their abdomen would feel those effects over a period of time and very likely that within ten years or so they would develop serious blood disease as a result of radiation. But the patients who would receive this heart were on the edge of death and it was certainly conceivable that they would take the risk of developing leukemia ten years in the future in order that their lives be saved immediately.

But in addition, regardless of how well shielded that capsule was, there were discernable risks to other parties, persons in close contact with the recipient of the artificial heart. It was difficult to quantify those risks and our committee asked a number of renowned experts around the country to help us understand what those risks might be. I remember vividly that we received back four very thick studies done by experts at Cal Tech and MIT and so forth, with rads and curies all calculated out. I have to admit that I did not understand much of those highly technical data. But from one expert came back a one page letter with one paragraph typed on it. It said, "My only concern about the totally implantable artificial heart is that some day I will find myself on a transatlantic flight sitting between two of them." I think that letter had a greater impact on me than the long studies.

But the bottom line of the studies was that there was a discernible risk to parties other than the recipient and that therefore the clinical use of this device would clearly have to entail certain regulation. The bearers would probably have to be identified in some way as carriers of a radioactive device. They would probably have to be licensed. They would probably have to be restricted from certain activities, such as those that brought them into close contact with children or other vulnerable population segments (for example, schoolteaching, nursing, pediatrics, and many others). There would have to be methods to recover the plutonium after their death, and so forth and so on. The problem posed to the committee was, should we think approvingly of a device which, if effective, would save the life of an individual but, at the same time, make that individual a walking risk in society. I believe that was probably the first time in history that one could seriously say medicine which was curative for an individual posed a risk to society.

The committee was not very enthusiastic about the totally implantable artificial heart with the nuclear power source. We spoke somewhat disparagingly of it and emphasized its dangers. Partially as a result of our report (and partially as a result of other factors, such as money, politics and scientific uncertainties) that particular approach to the power source was phased out. It has not been followed up subsequently by any federally funded research. However, continued research on the development of the blood pump and on other power sources, such as electrical and thermal batteries continued with the approval of the committee (The Artificial Heart Association Panel, 1973).

Some ten years after that first committee experience, I received another phone call from the new director of the National Heart, Lung and Blood Institute, Dr. Claude Lenfant. He said he was reconstituting the committee to review the current status of the artificial heart. There had been important new developments in the intervening decade between the first and the second reviews. The updated idea for a power source involved an internal battery and an external connection whereby the battery could be recharged. This second review examined not only the total heart device, but also several other possibilities, including an assist device. This device would not replace the heart, but rather would be implanted near the heart and assist it in its action, primarily by giving power to the pumping action of the left ventricle. So when the second committee met, we discussed not only the artificial heart but also mechanical circulatory assist devices, which included the total heart and partial hearts and external power sources. Two months after the second committee held its first meeting, in September, 1983, Dr. William De Vries implanted an artificial heart in the chest of Dr. Barney Clark of Salt Lake City, Utah (Working Group on Mechanical Circulatory Support, 1985; Shaw, 1984).

One of the important questions that must be asked in assessing the benefits of a new medical technology is how extensively will it be used? Several estimates of the extent of use of the artificial heart had been made in previous years. Our committee, with more sophisticated methods, judged that approximately 17,000 to 35,000 persons in the United States annually could benefit from an artificial circulatory assist device. These are patients who suffer from end-stage heart disease. End-stage heart disease arises either from very serious damage to the heart muscle that results from myocardial infarction or from other causes that create the condition called cardiomyopathy in which the heart muscle is so severely weakened that it loses its pumping power. Clearly there is a very serious health problem in the United States relative to cardiovascular disease. There are probably 100,000 deaths a year resulting from myocardial infarction. There are probably 50,000 deaths a year resulting from other forms of cardiomyopathy. It appeared to our committee that, on the basis of certain criteria for suitability, that somewhere between 17,000 to 35,000 of those deaths could be prevented if the damaged hearts could be replaced by either partial or total assist devices.

How much would it cost to do that? The cost would be calculated by multiplying the number of patients by the cost of the device and its implantation, together with the cost of any subsequent medical care over the expected life of the patient. It was difficult to predict the expected life of the patient with a mechanical device, but on the basis of fairly sophisticated projections, the best guess was that a patient's life, which would otherwise end very rapidly, would be extended about three and a half years. One hundred and fifty thousand dollars per patient was the best estimate, both for the cost of the device itself and for the subsequent care of the patient.

This cost is not excessive when compared with other forms of medical care for serious disease. Liver transplantation is more expensive, at about $200,000, bone marrow transplantation costs the same, about $150,000. One does not even have to look at these very exotic sorts of organ transplantation for comparable medical care costs. The care of a cancer patient can be in that range, $100,000–$150,000. The care of a burn patient is in the same approximate range. The care of a psychiatrically disturbed adolescent may range even higher.

So put within that context, the costs of many forms of medical care are comparable. Of course one must multiply the 17,000 to 35,000 patients who could use the device by the 150,000 dollars approximate cost per patient. This results in costs nationally in the range of two and a half to five billion dollars annually. When you see a figure of that magnitude you are inclined to ask, "Well, who is going to pay the bill? What effect will this device and its costs have on other forms of health care?"

In the United States with our form of health care, the answer to that is fairly easy. The patient will pay. Of course, in the United States, the majority of persons are covered by insurance. Therefore when we say the patient will pay we suggest that the insurer will pay. Clearly the insurer has to be satisfied that the device is not experimental or investigational, and that it can be considered a safe and effective standard practice (at least for some conditions). However, if you say the insurers will pay, what you are implying is that the consumers will pay. That is, everyone who buys insurance will eventually pay. The risks and the costs have been shared over the entire pool of individuals who buy insurance or are participants in a plan.

However, there is also another response to the question "Who will pay?" The government will pay. Because the government, under certain circumstances, does pay part of our health bill. When does the government pay? Since 1966 the government has paid for a portion of the health care of all individuals in the United States over sixty-five through the Medicare program. This now amounts to some forty-five billion dollars per year. In addition, since 1966, the federal government, in conjunction with the states, has paid for the health care of individuals who fall below the "poverty line," now somewhere in the range of eleven thousand dollars a year. Each state sets a level of percentage below the poverty line that it is willing to pay. The Medicaid program comes to some thirty-five billion dollars a year annually. So the government does pay for health care under those two circumstances – persons below a certain income level and persons over sixty-five.

However, the government also pays in another important situation. Specifically, it pays for patients who suffer from end-stage renal (kidney) disease. In 1972, the Congress was persuaded to amend the Social Security Act and the Medicare section in order to pay for the treatment of this one disease. There were two modes of care which were life-saving: renal dialysis and renal transplantation. The Congress was persuaded to pick up the bill for those patients and those patients only. In the United States we have a very peculiar form of national health insurance: we insure by the organ affected. If your kidneys go out you are covered; if it happens to be your stomach or your liver, or if you have cancer or a brain tumor, you are out of luck.

I raise the issue of the end-stage renal disease program for a very special reason. First of all, the Congress was persuaded to pass that special provision because it was made clear to them that this was a dramatic, life-saving intervention: without it patients would die rapidly. In so doing, they appealed to a deeply embedded American ethos – the rescue ethic. They did so in a dramatic way. The American Kidney Society actually dialyzed a patient in front of the Congressional hearing, and said, "If this man was unable to obtain this procedure, if he couldn't afford it, he would die immediately." Almost without question the Congress passed the end-stage renal disease program.

Today the end-stage renal disease program covers some 110,000 patients. It costs approximately the same amount that the artificial heart or the assist device would cost at the lower end of the estimate, about two billion dollars a year. Now the interesting point to note, however, is that two billion is approximately five percent of the entire Medicare expenditure. It serves less than one half of one percent of recipients of Medicare funds. So a large part of the total Medicare budget, which is otherwise targeted primarily to persons over sixty-five, is funnelled into the care of a very small group of persons, those with end-stage renal disease. While 110,000 seems like a lot of people, in terms of the large number of people who are recipients of Medicare funds, it is actually very few. This should make us wonder whether, if the totally implantable artificial heart were brought to a point of clinical use, there would not also be strong pressure to add that form of care as a special provision as well, adding another two billion dollars for another relatively small segment of the population who are in need of health care in general (Gutman, 1988).

Thus, the problem that is caused by the addition of these specific technologies to our national health care expenditures is whether the addition of high cost rescue technologies may not fulfill a life-saving ethic at the expense of another genuinely important aspect of modern medicine, namely life-enhancing care. Let me illustrate what I mean by example. In the state of Oregon in 1989, a momentous decision was made. The legislature of that state determined that the approximately one million dollars of state funds which would go towards the funding of organ transplantation for Oregon citizens would instead be devoted entirely to the support of programs in prenatal care. This decision by the legislature said, "The state of Oregon will no longer pay for any organ transplantation." The number of those transplantations each year was small. There were ten to twelve a year, done for persons who otherwise could not pay for transplantation. But that one million dollars, it was claimed, could serve approximately five thousand pregnant women to ensure that they had appropriate prenatal care. Prenatal care clearly has definite benefits for the health of child and mother, and many of the unfortunate problems that can arise due to prematurity and so forth, can be avoided by appropriate prenatal care.

Thus far, the state of Oregon is the only state in the union that has faced up to the problem of the rationing of health care explicitly. However, one year later, they began backing off because there was a great outcry in individual cases. One three year old, who needed a bone marrow transplantation, could not obtain it without state aid, but the state had made a policy decision and would not pay for the transplant. The child's mother, whose concern was about her baby rather than the state budget and policy, brought her baby to the neighboring state of Washington and said, "Help me. Help my baby." Clearly the state of Washington (or the State of California) has not been particularly happy about Oregon's bold decision. But the rationale behind that bold decision was an affirmation of what might be called a life enhancing ethic, rather than a life saving or rescue ethic (Anonymous, 1990).

Most of the children who benefit from prenatal care are not children who are about to die; they are children who are born with significant deficiencies and live with them throughout life. So the prevention of those deficiencies resulting from prematurity and other problems, is life enhancing.

Another obvious example of this is the question of smoking education. Our second committee at the National Heart, Lung and Blood Institute pointed out in its report that a very large portion of the cardiovascular problems that come to the point of requiring heart transplantation or implantation are the result of smoking. We are all aware of the enormous impact on American health of smoking addiction. It is estimated that some 800,000 deaths per year are in some way closely related to smoking addiction. The committee simply wondered, without going any further, whether or not 2 billion dollars devoted to helping people overcome an addiction, might not bear much greater fruit for the public health than the development of the totally implantable artificial heart or the circulatory assist devices.

Similarly, we are now quite concerned about problems of obtaining appropriate long term care, particularly for elderly persons and especially for those elderly persons who suffer from devastating forms of brain disease, such as Alzheimers disease, and other types of dementia. Again, those costs are significant. It is a major health problem in the United States and a major financial burden laid upon families. The committee wondered whether the funds which would otherwise go to the rescue ethic of the artificial heart, might be better spent in that direction. It would be a small but important feature of American health care to attempt to remedy that problem. Others that might be mentioned are rehabilitation after stroke or after accident or trauma, or the attention to various musculoskeletal diseases such as arthritis, which is the major debilitating disease that keeps people away from work in the United States. One could identify many other areas in which significantly deficient attention is paid and money spent on what could be considered life enhancing activities.

Now the 2 billion dollars added to the artificial heart is not going to make a big dent in any of those, but as one adds 2 billion dollar increments for high technology devices the squeeze has got to come somewhere. It is common enough to hear people respond to this problem by saying that the money should come from the defense budget. This of course, is an expression of the respondent 's values and may be widely shared. But it is a hopelessly naive view of the federal budgeting process and politics. Federal health budgets are finite, not open ended. Medicare, Medicaid, the public health service, the Veterans Administration, all of these work with specific dollar amounts determined by the Congress. The addition of charges to those budgets can be accommodated either by adding new money or squeezing out old money. New money in large amounts is highly unlikely in our present political and fiscal situation and, even in more favorable circumstances, all new money must be raised somewhere and compete with other legitimate interests. Thus, the most likely reaction to a new charge is a squeeze on existing resources. It is fairly obvious that it is much easier to curtail and trim the budget of a less visible and less dramatic form of care. If, for example, one has pre natal clinics, and you have a budget squeeze, it is fairly easy to close the clinic an hour earlier, and say, "instead of staying open until eight o 'clock we 'll only go until seven o 'clock. Instead of having six nurses, we 'll have five nurses." It is possible to save a lot of money that way. One can implement that sort of program cut with relative ease and hardly anybody notices it except the person who gets off work at six-thirty, except the people who receive the service and have to wait much longer, who have to travel further, or who find that the clinic in their area has been closed. From the administrator 's point of view, that 's an easy saving. That sort of squeezing can go through the entire realm of services. Because these forms of care are less visible, they are easier to limit by small reductions. They are also less attractive to institutions and to providers because often they bring in much lower reimbursement from insurers and from the government.

The paradox that we face, then, when looking at these technologies is this: we simply have no assurance that money that is "saved" by not developing a device like the totally implantable artificial heart will go to more effective health care programs. That is not the way federal budgets work, although occasionally there is pressure to make a change of that sort within a very specific situation and that happened recently. In 1988, Dr. Lenfant, Director of the Heart, Lung and Blood Institute, tried to shift twenty-two million dollars within the circulatory assist budget from the total heart program into the left ventricular assist program. He cancelled several contracts that were devoted to the continued development of the total heart and put it into the development of the left ventricular assist devices which he and many others believe to be a much more useful technology. Such a change is the prerogative of a director of an institute. The next day, however, the papers reported that Senator Kennedy and Senator Hatch (who by the way comes from the state of Utah, which had one of the big contracts for the totally implantable artificial heart) had called Dr. Lenfant on the carpet and said that if that money was not restored that he would see significant cuts in the budget of the Heart Institute. The ability to make those sorts of transfers is limited. Even when it happens, the results are not always those desired and demonstrate the highly political nature of the budget process (Anonymous, 1988).

There is another problem that makes any sort of rational planning for allocation of medical resources extremely difficult in the United States. In the United States no single entity does the developing and marketing and provision of health care. For example, the federal artificial heart program was largely an inhouse program at the National Institutes of Health. The money was spent by funding scientists working there and by letting contracts to scientists elsewhere. However, at the same time, a commercial development was underway that was not dependent on federal money. The Jarvik heart, used for Barney Clark and several other recipients, was developed largely by commercial financing. In fact, in a letter to the New York Times, following the attempted budget shifts in the artificial heart program, Robert Jarvik said that he believed it was time for a new initiative toward development of the artificial heart through the private sector (Jarvik, 1988). The problem is that if the government makes a decision to discontinue or to shift funding, commercial developers who see some promise for profit in a particular development can certainly go ahead. Indeed at the time our first committee met in 1972, we were informed that the United States had to move ahead rapidly with the artificial heart because Toyota was moving very rapidly toward the development of theirs and they would be marketing that heart in the United States. If they could not get FDA approval here, there would have been nothing (except poverty) to prevent people with end-stage cardiac disease from flying to Japan for a Toyota heart.

The development of a device like the totally implantable artificial heart or the circulatory assist devices have the effect of expanding inequity, particularly if such technologies are developed by commercial sources. Those who can pay will have their lives saved, those who cannot pay will die unless the government pays the bill. I said these technologies pose a problem of expanding inequity and I simply illustrated that by saying people who have the money to buy a form of care can buy it and those who can not lose out. But the problem of inequity in health care is an even broader and more difficult one. It is very difficult to think about equity in health care. It is now a favorite subject of people who work in medical ethics. A lot is being written about it. There have been some interesting and useful and illuminating books such as Norman Daniels's *Just Health Care* and Dan Callahan's *Setting Limits* (Daniels, 1985; Callahan, 1987). Nevertheless it remains very difficult to think clearly about the implications of rationing health care. When one thinks about equity in health care, it is necessary to consider the enormous range of factors that are involved. The term "health care" covers a vast set of different activities and institutions and processes and products, carried out by vast numbers of different providers with different training and background and interests. In addition, health care has differential outcomes, many of which are almost impossible to evaluate. All of those things somehow have to be comprehended in order to begin to make sense of what it means to talk about equity in health care. Certainly equity does not mean each person shall have an artificial heart. One could say that what equity means is that somewhere in the country there will be a storehouse with an artificial heart waiting for each of us. Obviously, that is stupid. Equity could mean that whatever happens to an individual by way of a health problem, each person will be able to have a remedy and that somebody is responsible for providing such a remedy. We each have a claim on a remedy. I do not have to elaborate to make it clear what an enormously complex and enormously burdensome result that would have. Equity certainly can not mean that.

What then can equity in health care mean? At least one thing that it might mean is that all citizens should share equally in risk. Then the question arises: risk of what? What are we sharing the risk of? Is it the risk of death? I leave you, the reader, to ruminate on that. I would suggest that the risks that we share are the risks of disability if not death. Disability means living with a reduced capacity for maintenance of one's self, one's life, one's relationships, and one's finances. We all share the risk of being the subject of some sort of damage to our body that makes it difficult to function. There are, of course, an infinite number of events that can render a person disabled. Among these possibilities, we might consider stroke; some six hundred thousand people suffer stroke followed by some neurological damage that leaves them with serious disability. It is important that they have readily available forms of rehabilitation which is oftentimes quite effective in restoring function, mobility, speech, and so on. The same thing is true of myocardial infarction, the same thing is true of trauma and accident. If one thinks of sharing the risk of disability, one can suggest that it is more imperative to repair damage, to restore ability, to prevent events to the extent possible, rather than to rescue from imminent death.

Let me end this chapter by relating an event that happened to me recently. Three visitors from Beijing Medical University came to Seattle to visit my ethics program at University of Washington School of Medicine. They were professors at the largest medical university in China. They wanted to know what we in the United States considered problems of medical ethics. So I took them to the places where our medical ethics problems arise. I took them to the adult intensive care unit, to the cardiac critical care unit, to the pediatric intensive care unit, and to the neo-natal intensive care units. Those places are the seed beds of ethical problems in the United States: the places where high technology medicine is doing both great good and oftentimes no good at all.

In the pediatric intensive care unit we saw a baby, almost three years old, who had a very peculiar disease. This baby was dependent upon a respirator. There was no chance that it would ever be able to breathe on its own. It was also blind and entirely deaf. At best, because of its condition, it may survive another year. By this point its bill was already half a million dollars. My three Chinese visitors looked at the baby and one of them said, "I could immunize a million children at the same cost." That was a very sobering remark, and points to one aspect of the ethical problem. Another event that happened recently illustrates another perspective. A very close friend of mine was a leading pediatrician and scientist. At age 57, he died quite suddenly of heart failure due to cardiomyopathy. He probably could have been saved by a heart transplant or an artificial heart implant. My friend's care might have cost $150,000 and, if we had a national program of insurance for heart transplants and implants, care of persons like him would cost two billion dollars. Yet, what benefits would his saved life have brought to children? (He was, peculiarly enough for this comparison, an expert in infectious diseases and immunization.) How many children would his life, his saved life, also save? That to me is a paradox. What is the value of the saving of one life, such as the life of this pediatrician, which may have had a ripple effect for the benefit of many thousands of children? What is the trade-off that is achieved by not expending the money for that artificial heart?

I will not attempt to discuss this further, primarily because of my own perplexity in the face of this problem of equity in health care. It is my belief, although I have a hard time constructing an airtight logical argument in support of this belief, that one of the features that will enhance equity rather than produce inequity is a great caution about the introduction of expensive life-saving technologies. That has been the burden of my remarks in this chapter. Yet, I still find myself troubled by the paradox.

References

Anonymous. 1988. "Federal Agency in Shift to Back Artificial Heart." *The New York Times*, July 1, 1988.

Anonymous. 1990. "Oregon Lists Illnesses by Priority," *The New York Times*, May 3, 1990.

Callahan, D. 1987. *Setting Limits*. New York: Simon and Shuster.

Daniels, N. 1985. *Just Health Care*. Cambridge and New York: Cambridge University Press.

Gutman, R.A. 1988. "High Cost Prolongation: The Kidney Dialysis and Transplantation Study." *Annals of Internal Medicine* 108: 896–899.

Jarvik, R. 1988. "Revive the Artificial Heart with Money and Vision," *The New York Times*, August 8, 1988.

Jones, W.H.S., trans., 1923. *Hippocrates, Volume 1*. London: William Heinemann.

Jonsen, A. 1973. "The Totally Implantable Artificial Heart." *Hastings Center Report* 3: 1–4.

Shaw, M.W. 1984. *After Barney Clark*. Austin: University of Texas Press.

The Artificial Heart Assessment Panel. 1973. *The Totally Implantable Artificial Heart: Legal, Social, Ethical, Medical, Economic, Psychological Implications*. National Heart and Lung Institute. Washington, D.C: Department of Health, Education and Welfare.

Working Group on Mechanical Circulatory Support. 1985. *Artificial Heart and Assis Devices: Directions, Needs, Costs, Societal and Ethical Issues*. Washington, D.C.: National Heart Lung and Blood Institute.

Chapter 5

Hazards of the American Health Care System: No Treatment, Under-Treatment, and Over-Treatment

DEBORAH MATHIEU

Although the vast majority of Americans are covered by some form of health care insurance, the coverage is incomplete. It usually involves some form of co-payment and/or deductible on the part of the patient; it does not include all beneficial health care goods and services; and it does not include all individuals. Most people are well aware of some of the dire consequences of the large gaps in the patchwork of insurance coverage, for the media frequently portray stories about patients who seem to have been left out of the health care system: an AIDS patient who cannot afford the latest drug, an end-stage cardiac disease patient who cannot pay for a heart transplant, a leukemia patient who cannot afford the search for a compatible bone marrow donor.

But while these accounts make an important point about the obstacles confronting some individuals in obtaining beneficial medical care, they also may be misleading. By concentrating on those few diseases the media regard as especially newsworthy, the reports give the false impression that persons with other, less controversial or more common conditions have no difficulty in securing health care services. And this is simply not the case. A significant portion of the American population, including a large number of children, cannot afford and do not have access to adequate health care.

Approximately 37 million Americans have no health insurance at all – 11 million more than a decade ago – and almost forty percent of these are children (Emmott and Wiebe, 1989). The unemployed are not predominant among those who have no insurance. Approximately sixty percent of uninsured adults are employed but receive no coverage through the workplace. Not surprisingly, lack of insurance is a greater problem for low-wage workers – blue collar, service, and agricultural – than for white collar workers.[1] Many of these people cannot afford to buy insurance on their own, yet they earn too much money to be eligible for Medicaid (the federal program that pays for the medical care of the poor and the medically indigent). A family of four in Alabama whose total annual income is $4,500, for instance, would earn too much to be eligible for Medicaid benefits in that state.[2]

Lack of health care insurance often coincides with lack of access to health care services. Studies show, for example, that people who do not have health

M. Waterstone (ed.), Risk and Society: The Interaction of Science, Technology and Public Policy, 75–89.

insurance often do not seek health care when they are sick. A 1986 survey sponsored by The Robert Wood Johnson Foundation found that people who did not have health insurance were likely to forgo medical care, even when faced with symptoms – such as chest pain or unexplained bleeding – which may indicate life-threatening conditions.[3]

And some of the uninsured who seek medical care, even life-saving care, do not receive it. Despite a federal law requiring health care institutions to provide life-saving emergency treatment to anyone who needs it (regardless of the patient's financial situation), many hospitals continue to turn away patients who cannot pay for their care. The President's Commission for the Study of Ethical Issues in Medicine and Biomedical and Behavioral Research (hereafter, President's Commission) reported in its 1983 study of access to health care that there are wide disparities of access based on inability to pay and the amount of health care coverage (President's Commission, 1983), and the National Leadership Commission on Health Care came to a similar conclusion in its 1989 report (Executive Summary, 1989).[4]

Improving Access to Health Care

While unfortunate, this inequality of access to health care is not illegal. There is no general legal right to health care in the United States (as there is in most other Western industrialized nations). There is no law mandating that medical benefits are offered to everyone regardless of ability to pay; and the Supreme Court has ruled that even though Congress has opted to subsidize many medically necessary services through various programs, it has no legal obligation to subsidize all medically necessary services.[5]

Yet there is a strong feeling in this country that health care is not merely another commodity to be bought and sold on the market, and that it has a special importance in our lives that mere commodities do not. The President's Commission, which documented the persistence and prevalence of this view in American public life, has argued that the special moral importance of health care creates societal obligations to distribute it in ways that might not coincide with the results of an unfettered market distribution:

> Society has an ethical obligation to ensure equitable access to health care for all. This obligation rests on the special importance of health care: its role in relieving suffering, preventing premature death, restoring functioning, increasing opportunity, providing information about an individual's condition, and giving evidence of mutual empathy and compassion. Furthermore, although life-style and the environment can affect health status, differences in the need for health care are for the most part undeserved (President's Commission, 1983).

And the National Leadership Commission on Health Care (Executive Summary, 1989) concluded that, "There should be no financial barrier separating Americans in need of health care from access to available care."

Need for health care, then, not ability to pay, is often championed as the relevant distributive criterion for health care. On this view, our health care system – where a significant portion of the population cannot afford and often therefore does not receive adequate medical care – is morally suspect.

Many people believe that this moral stance should be reinterpreted in the form of a legal right to health care. A 1987 Harris poll, for instance, indicated that over 90 percent of the respondents agreed with the claim that, "everybody should have the right to get the best possible health care – as good as the treatment a millionaire gets." If adopted as public policy, however, this position could lead to some unappealing results. If the level of health care to which everyone is entitled were set as high as technically feasible, then the commitment to providing everyone with the very best and most expensive health care would place an unbearable burden on social resources. To dot the Alaskan tundra and the Arizona desert with high-technology hospitals, for instance, would be extremely costly and inefficient. Guaranteeing the same high level of health care to everyone may require so many resources that other important social goods would have to be eliminated.

Since resources are finite and health is not the only good, the argument for improved access to health care is generally considered to be most forceful when applied, not to all available health care goods and services, but to an important subset of them. The most reasonable idea, then, is not to try to meet everyone's demands – that would be too costly – but to provide access for everyone to at least a basic level of health care.

The institution of a national health insurance program to improve access to health care in the United States is slowly becoming a real possibility. The idea is not new – Theodore Roosevelt used it as part of his campaign for President, and it has been resurrected periodically ever since – but the current level of interest in the idea is unprecedented (see Bowler et al., 1977; Starr, 1982). Even long-standing foes of national health insurance – such as physicians' groups and large corporations – are now considering the idea favorably.[6] For the first time in its history, for instance, the American Medical Association gave its support in June 1989 to the concept of a federally mandated, employer-provided health insurance.

While there is much disagreement about which services would be included, some benefits are uncontroversial. One of these is prenatal care, which is effective in reducing infant mortality and morbidity (see Gortmaker, 1979; Fisher et al., 1985; Rahbar et al., 1985; Moore et al., 1986; Murray and Bernfield, 1988). Prenatal care has been shown to be cost-effective as well.[7] But improving access to this beneficial service is more problematic than it may first appear.

Although the vast majority of pregnant women already have access to prenatal care, there is a small and worrisome contingent who receive no prenatal care at all or medical attention only during the last trimester. Those most likely to receive little or no prenatal care are indigent, unmarried black teenagers – and these are precisely the individuals who are at greatest risk of having low birthweight babies (Institute of Medicine, 1985). One result of this lack of prenatal care is that many children are born handicapped who could have been born healthy. In one recent case, a woman who had not received prenatal care before delivery bore a child suffering from cataracts as well as liver and heart problems caused by her syphilis – all of which could have been prevented had she been given a $20 shot of penicillin. The child survived for only a month (at a cost of over $70,000) (Nazario, 1988).

It is naive to think, however, that making affordable medical care available to all pregnant women would solve the problem of access to prenatal care. Many pregnant women do not receive medical care because they do not seek it. These women are stymied by a lack of transportation, or an inability to find someone to care for their other children, or a fear of the medical establishment, or a lack of knowledge about the importance of prenatal care and/or the availability of services, or an unwillingness to cope with the inconveniences and long waits at many clinics (Klein, 1971; Joyce et al., 1983; President's Commission, 1983; Poland et al., 1987). Simply offering health care services, therefore, would fall short of the goal of substantially improving the health and increasing the chances for survival of these infants. Supplementary public education programs are also needed, as are incentive programs to encourage pregnant women to seek care. Money could be offered for each prenatal visit, for example, or services – such as transportation to the health care institution and child care – could be provided.

We should not be too quick, in other words, to see "health care" only in terms of "medical care" (as those who design proposals for a national health insurance program tend to do). Although making medical care universally affordable is a step in the right direction, it will not guarantee a healthier population. The most competent physician in the world cannot cure a patient who does not seek treatment. Hence it is important to offer other related forms of assistance and incentive programs in order to reach those whose barriers to medical care are not entirely financial.[8]

Other Allocational Issues: Under-Treatment and Over-Treatment

Being kept out of the health care system is only one hazard facing individuals who require medical care. Other less obvious but equally undesirable risks await those who are accepted for treatment. There is growing evidence, for instance, that treatment decisions are sometimes made for irrelevant or biased reasons. Put plainly, some patients are victims of unfair discrimination. This discrimination comes in two forms: some people do not get enough medical care, while others get too much. These problems will not be solved by throwing more money into the pot, for financial barriers to care are not really the issue.

Under-Treatment

More and more studies indicate that some medical treatment decisions are influenced by unfair (albeit perhaps unconscious) bias, and that beneficial care is withheld from patients as a result. A striking case study involves patients with end-stage renal disease (ESRD).

In 1972, end-stage renal disease (the permanent failure of the kidneys to collect and dispose of bodily wastes) became a very special condition (for reasons discussed by Jonsen in Chapter 4 of this volume), for in September of that year Congress decided to expand the Medicare program to cover the medical expenses of virtually all individuals with that life-threatening disease. Medicare had already been paying for some end-stage renal disease care (patients who were eligible for Medicare benefits could receive treatment for ESRD) but in 1972 Congress widened Medicare's eligibility criteria to include almost all sufferers of this condition, regardless of their age, and thereby greatly widened access to the expensive life-prolonging technologies of artificial kidneys and transplantation.[9]

Although patients with ESRD have equal access to treatment by artificial kidney, this does not seem to be the case with regard to kidney transplantation. There is evidence of unequal and suspect selection of patients for transplantation. In Washington, D.C., for instance, approximately 22 percent of the transplants performed using cadaver kidneys in 1985 involved foreign national patients. This rate was considerably higher than the national rate for kidney transplantation into foreign nationals, which was about 5 percent. But it is especially noteworthy for two reasons: 1) a major military hospital located in the District of Columbia refused to accept additions to its waiting list for organ transplants because of the shortage of organs; and 2) the incidence of patients on dialysis in Washington is more than double the national average (Office of the Inspector General, 1986). One concern here, of course, is that foreign nationals are getting organs that could have gone to U.S. citizens. But another concern is that the majority of people who lost out are those of a particular minority group, African-Americans. About 70 percent of the Washington, D.C. population is African-American, and the rate of end-stage renal disease among this population is four times higher than among whites in the United States.

In general, African-Americans tend to receive proportionately fewer kidney transplants than whites. Indeed, a study of kidney transplants performed nationwide in 1986 showed that the typical transplant patient today is a white male with a good income – and this is precisely the profile of the typical transplant patient before the federal government began its ESRD program. The percentages have changed since the early 1970's – more women, lower-income patients, and minorities are receiving transplants now – but whites are almost twice as likely as blacks to receive kidney transplants; men are about one-third more likely to receive transplants than women; and higher income patients are more likely to receive transplants than lower income ones (Held et al., 1988).

The reasons for this differential treatment are enormously complex and may have little to do with unfair discrimination. African-Americans, for instance, are less likely to want a transplant than are whites, and African-Americans do not do quite as well as whites after transplantation (Callender, 1986). In addition, the immunosuppressant drugs required by transplant recipients are very expensive – about $6,000 a year – and patients bore the entire financial burden for them during the periods of the studies cited.[10]

Nonetheless, the relatively lower rate of kidney transplantation in African-Americans, women, and low-income ESRD patients is a cause for concern, and certainly deserves closer scrutiny. The situation is especially troubling because it appears to be part of a pattern which extends throughout the health care system. A study of patients with heart problems in Massachusetts, for instance, concluded that "race does seem to be important" in treatment decisions.[11]

Citing a study published in 1981 by the Institute of Medicine, the President's Commission reported that "there is considerable evidence that racial and ethnic factors are associated with disparities in patterns of health care and clearly support the concern of many people that minority groups are still discriminated against in this country." The President's Commission itself told about reports it had received concerning members of minority groups who were seriously ill or badly injured, as well as women in active labor, who were turned away from hospitals, transferred to other (public) hospitals, or subjected to long delays before they received care (President's Commission, 1983). And authors of a more recent nationwide study of access to health care concluded that,

> [T]here continues to be a lack of parity in access to health care, and a consequent excess of unmet needs for blacks compared with whites. The difficult economic circumstances of many black families clearly contribute to the lack of access to health services. However, the findings suggest that even blacks above the poverty line have less access to medical care than their white counterparts. Despite progress during the past two decades, the nation still has a long way to go in achieving equitable access to health care for all its citizens (Blendon et al., 1989).

This seems to be true in the realm of medical research as well. A literature review of drug studies conducted over a three-year period, for instance, concluded that blacks are under-represented in clinical trials of new experimental drugs, even when the drugs are designed to treat ailments suffered more by African-Americans than by whites (such as sickle-cell anemia) (Department of Health and Human Services, 1985; Svensson, 1989).

Other indications of inequitable treatment were uncovered by researchers who reviewed the hospital charts of over 1,400 patients with a diagnosis of non-small-cell lung cancer in New Hampshire and Vermont between 1973 and 1976. Their conclusion was that the patient's condition did not always account for the medical treatment received, but this time the patient's race was irrelevant. Instead, "potentially curative or palliative treatments were not provided to patients who seemed less able to pay or who lacked a spouse" (Greenberg et al., 1988). It is easy to understand why ability to pay would make a difference; but what about marital status? Perhaps marital status was irrelevant to the treatment decisions and it just happens to coincide with some other, more telling (but as yet unidentified) marker. We cannot be certain that prejudice was an element; but neither can we be confident that it was not.

It is alarming to suspect that, even if financial barriers to health care were removed, other more insidious barriers to adequate care would remain for some Americans.

Over-Treatment

Researchers who found that some patient groups tended to get less treatment than others noted that this could indicate either under-treatment of one group *or* over-treatment of another (or both) (Weissman and Epstein, 1989; Wenneker and Epstein, 1989). Determining which is the case is clearly the next step. In the meantime, however, indications of over-treatment raise serious concern.

A study of cardiac bypass procedures performed in three hospitals, for instance, concluded that 44 percent of the operations were questionable, and that 14 percent of them were inappropriate (Randal, 1982; Anonymous, 1984). And researchers who investigated the use of carotid endarterectomy to remove a clot from a neck artery concluded that only one-third of the 1,302 procedures performed were beneficial (Ludtke, 1989). Similar problems exist with respect to the pacemaker (a small battery-operated device which emits electrical impulses to correct heart irregularities). Implanting these devices is a very common procedure – surgeons in the U.S. implant approximately 100,000 each year – but there is increasing evidence that the procedure is *too* common. Researchers who reviewed pacemaker implantations in thirty hospitals in Philadelphia concluded that twenty percent of the implants were not medically indicated, and that thirty-six percent were not adequately justified (Greenspan et al., 1988). If this were true of pacemaker implantations nationwide, then there are at least 20,000, and perhaps as many as 56,000, unwarranted operations each year.

It is ironic that while many people in this country receive too little medical treatment, many others receive too much. One explanation for this apparent trend of over-treatment is that many physicians overestimate the therapeutic value of the procedures they recommend. The Office of Technology Assessment (OTA) has estimated that only between 10 and 20 percent of all medical techniques have been demonstrated to be efficacious by controlled clinical trials (Office of Technology Assessment, 1984). It should come as no surprise, then, that studies will reveal that some common surgical techniques are appropriate in fewer situations than formerly believed. Further research into patient outcomes and widespread dissemination of the results should go far to improve the situation.[12]

But this is only one part of the story. There are other influences at play here as well, which can be illustrated by using the example of Caesarean sections.

Between 1968 and 1977, the Caesarean section rate in the U.S. tripled; by 1978, over 15 percent of all births were by Caesarean section, making it one of the most common surgical procedures performed. In 1980, a consensus was reached at a national conference that this rate was much too high, and the participants recommended that the trend be stopped (Anonymous, 1981). But that has not happened. In fact, the very year after the national consensus conference published its report, the national rate had climbed to almost 18 percent (Gleicher, 1984), and by 1985, almost twenty percent of all deliveries were by Caesarean section (Teffer et al., 1987).[13] Many of these operations are unnecessary, and many of them pose greater risk to the mother than normal delivery: the mortality rate is two to four times higher for Caesarean sections than for vaginal deliveries (Gleicher, 1984).

Why are so many Caesarean sections performed? One answer is that women often prefer the procedure to vaginal delivery, believing that it will improve their chances for having a healthy child. But it is also important to examine the factors which induce physicians to comply with these requests. One factor is fear of legal liability. Women whose babies are imperfect when born often blame the obstetrician and sue him or her. Another is convenience. The physician can schedule a Caesarean section at his or her convenience, while the baby determines when it will be born vaginally (and this may be very inconvenient indeed). There also are strong financial incentives to the doctors and the hospitals. Physicians charge more – and insurance companies are willing to pay more – for Caesarean sections, even though Caesarean sections take much less time than regular deliveries (which may take many hours). The hospitals also benefit because women and their babies stay in the hospital longer after a Caesarean section, and so they (or their insurance company) pay more.

In light of this, it is interesting to note which women are most likely to receive Caesarean sections. The women who receive Caesarean sections most frequently are the women with the most comprehensive insurance coverage, and those who receive Caesarean sections least frequently are those who have no health insurance (Placek et al., 1983; Williams and Chen, 1983; Gould et al., 1989). An alarming factor here is that the women who cannot afford the procedure are usually the ones who need it the most. Poor women tend to be less healthy and more at risk of giving birth to an impaired or low birthweight baby than affluent women (Institute of Medicine, 1985).

Implications and Conclusions

Individuals lacking financial resources have always encountered difficulties in obtaining adequate health care in our largely private, fee-for-service system, and the tragic consequences have not gone unnoticed (the story of an otherwise virtuous adult who takes to a life of crime in order to pay for the surgery needed to save the life of a parent or child was a favorite theme in early melodramas).

The federal government, in response, has long been in the business of improving access to health care; the Medicare and Medicaid programs are no doubt its best-known ventures.[14] In addition, state and local governments own hospitals whose main mission is to care for the poor; many physicians sometimes forgo their fees when a patient cannot pay; and individuals donate money to health care organizations which serve the needs of the poor. Yet despite these and other efforts, a significant number of people in this country continue to go without adequate health care and incur disproportionate health risks.

The move to effect a remedy has gained considerable political momentum recently, and no doubt access to health care will be widened to some extent. At this stage, however, where the large majority of Americans already have some level of health insurance, there needs to be a change in emphasis. Simply making payment mechanisms available for medical treatment (as in the Medicare and Medicaid programs) is no longer enough to improve significantly the level of health in the country. To widen access to health care we must also educate people about the value of that care and make it less onerous for them to obtain it.

In addition, it is important to note that sufficient resources to provide for all health care needs and preferences will never materialize – unless we are willing to permit serious constraints on our ability to produce other things of value – so questions regarding the proper allocation of health care goods and services will always exist. Because we will never be able to offer everything to everyone, choices about who gets what level of medical treatment are unavoidable. And these choices are open to abuse. We must be careful to differentiate between ethically acceptable inequalities of access to treatment – based on legitimate concerns for patient benefit and indications of "medical suitability" – and ethically objectionable inequalities of access. To guard against unfair bias, whether unconscious or deliberate, each criterion employed in the patient selection process must be scrutinized carefully. All require ethical justification as well as scientific support, and all should be examined for discriminatory effect.

Risk of under-treatment is a very serious problem; but the risk of over-treatment may be just as dangerous. One concern raised by attempts to widen access to the health care system is that it will exacerbate the tendency to overtreat paying patients.[15] Some of this apparent over-treatment may be due to lack of information, a deficiency which further research into the therapeutic value of various treatment modalities should help to remedy. But some of it is due to financial incentives, and health care providers are no more naturally immune to these incentives than anyone else.[16]

This trend must be stopped. As a society, we need to devote more attention to redesigning our health care system so that fewer treatment decisions are made for financial reasons. Not only would reducing the incidence of over-treatment save lives, but it would save valuable and limited resources as well. One element of reform would involve greater socialization to strengthen health care professionals' acceptance of professional mores which would counter-act the natural inclination to be influenced by economic considerations. Another would remove as many monetary inducements as possible (e.g., by requiring second opinions, putting physicians on salary, denying payment for laboratory work to physicians who send their patients to labs they own, and so on).[17]

In sum, it is important to recognize the hazards our health care system poses for those who seek help. Some would-be patients are shut out of the system altogether, while those who are accepted for treatment face the risk that decisions regarding their care may not be made in their best interests.[18] Thus before we get carried away with the idea of pouring more money into the health care system in order to improve access to it, we must first confront the daunting fact that it is all too easy and too common for medical treatment decisions to be made for less than laudable reasons: lack of information, bias, convenience, greed. While to a certain extent these problems can be charged to individual health care providers, to a much greater extent they are due to flaws in the health care system itself. Decisions made on the macro level configure the shortcomings on the micro level.

Will the United States ever have a health care system in which benefits and burdens are distributed both efficiently and fairly? Given certain facts of the current situation – especially the patch-work nature of the health care system (where no one is in charge and there is no coherent set of checks and balances), which is embedded in an over-arching social commitment to competition and the free market – there is good reason to believe that the only possibility for providing a fair and efficient allocation of health care goods and services lies in a fundamental revamping of our social institutions (see Baily, 1988; Daniels, 1988). Since such sweeping change is unlikely, and since discussion of this controversial issue is beyond the scope of this chapter, I will simply answer the last question in a word: No.

Notes

1. It is estimated that 24.2 million adult workers – primarily in the service, retail trade, construction, and agriculture-forestry-fishing industries – have no employer-group health insurance; seventy percent of these are employed full-time. (See Swartz, 1989).
2. States set their own eligibility limits for participation in Medicaid, and they vary widely.
3. See Anonymous, 1988.
4. Consider, for example, the case of 21-year old Terry Takewell. In September 1986, Takewell, an uninsured diabetic, was rushed by ambulance to a hospital near his home in Tennessee. After admitting him, hospital administrators discovered that he had an outstanding bill of $9,400, and quickly ushered him out of the hospital. Neighbors found him in the parking lot and brought him home; he died about twelve hours later. (See Ansberry, 1988).

   In documenting the differences in hospital care provided to uninsured patients compared to insured patients, researchers concluded that, "lower overall utilization by uninsured patients persists across hospital types and over a wide range of medical conditions." (See Weissman and Epstein, 1989; Sloan et al., 1986; Duncan and Kilpatrick, 1987).
5. Maher v. Roe, 432 U.S. 464, 1977; Harris v. McRae, 448 U.S. 297, 100 S.Ct. 2671, 1980; Webster v. Reproductive Health Services, 109 S. Ct. 3040, 1989.

6. Those who formerly opposed the concept of a national health insurance program have a variety of reasons to endorse it now: in addition to improving access to care, it is expected to check the runaway costs of the health care system (an important goal for businesses whose health insurance costs increase significantly every year), and to reduce the number of unpaid medical bills (an important goal for health care providers).
   (See Enthoven and Kronick, 1989; Himmelstein, et al., 1989; Relman, 1989; and Anonymous, 1989).
7. The Institute of Medicine calculated the cost of providing high quality prenatal care throughout pregnancy to women who have the highest risk of bearing low birthweight children (welfare recipients with less than a high school education), and measured this cost against the savings that would be gained (in hospitalization and ambulatory care costs) by a reduction in the incidence of low birthweight:

   > The analysis showed that if the expanded use of prenatal care reduced the low birthweight rate in the target group from 11.5 percent to only 10.76 percent, the increased expenditures for prenatal services would be approximately equal to a single year of cost savings in direct medical care expenditures for the low birthweight infants born to the target population. If the rate were reduced to 9 percent (the 1990 goal set by the Surgeon General for a maximum low birthweight rate among high-risk groups), every additional dollar spent for prenatal care within the target group would save $3.38 in the total cost of caring for low birthweight infants, because there would be fewer low birthweight infants requiring expensive medical care. (See Institute of Medicine, 1985: 17).

8. This was one of the main conclusions of the President's Commission in its 1983 report. Several cities have instituted generous prenatal care programs. In Washington, D.C., for instance, any pregnant woman whose annual family income is less than $20,000 can receive free prenatal care, and a van (the Maternity Outreach Mobile) is available to transport pregnant women to and from the medical clinics (as well as to remind them of their appointments).
9. The majority of ESRD patients are treated with an artificial kidney. The most common method is hemodialysis, in which an artificial kidney machine filters the patient's blood and then returns it to his or her body (usually three times a week). Another method is peritoneal dialysis, which is gaining in popularity since the development of continuous ambulatory peritoneal dialysis. Rather than being hooked to a machine, the patient continually dialyses herself by infusing dialysate from a flexible plastic container (which can be kept in a pocket) into her peritoneum.

   Although these methods prolong lives, they are not cures. In addition to continuing to suffer from the life-threatening disease that destroyed their kidneys, many dialysis patients are discomforted by the procedure. Most of them are tethered to machines for several hours, several days a week, and many of them experience weakness, lethargy, depression, loss of appetite, sleep disturbance, and frustration over their restrictive diet. And their condition remains life-threatening:

   > End-stage renal disease leaves no organ spared. It is a systemic illness that challenges the therapist to the utmost. Each attempt by physicians to intervene creates a new series of problems. Drug toxicity is rampant. While methods of dialysis ameliorate certain complications and afford symptomatic relief in some areas, the patient often becomes subjected to progressive bone disease, access-related infections, and insidious encephalopathologies (Stone, 1983).

   The other form of treatment for ESRD is kidney transplantation from either a living related donor or a cadaver donor. While the one-year survival rate for transplanted cadaver kidneys is approximately 71 percent and the one-year survival rate for donor kidneys is 88 percent, the one-year patient survival rate for kidney recipients is approximately 95 percent (the patient survives after the organ fails because he or she is put back on dialysis) (See Task Force, 1986).

   An organ transplant is not a cure, however, and an organ recipient is at constant risk of suffering recurrent infections as well as organ failure or organ rejection. In addition, the immunosuppressant drugs – which greatly improve a patient's chances for survival – also cause a variety of life-threatening conditions, including kidney and liver failure, lymphomas, and diabetes (See Starzl, 1983; Kahan et al., 1987).

Nonetheless, those ESRD patients who have had kidney transplants (and therefore no longer require dialysis) report that their quality of life has improved considerably (See Evans et al., 1985).

10. Medicare now pays for 80 percent of immunosuppressant drugs for one year after transplantation.
11. The researchers found that, compared to blacks, "whites underwent one third more coronary catheterizations and more than twice as many coronary bypass grafts and coronary angioplasties. These interracial inequalities were not merely a function of diminished physician contact and lower disease recognition for blacks. Rather, they were evident even among the cohort of individuals hospitalized for serious cardiovascular conditions." (See Wenneker and Epstein, 1989).
12. Fortunately, there is growing interest in the new field of patient-outcomes research, and health care institutions across the country are beginning to collect data on how well various treatments work. In addition, the Department of Health and Human Services (DHHS) is sponsoring a series of five-year studies of Medicare data to compare the effectiveness of different techniques to treat common medical problems (e.g., cataracts, heart attacks, and lower back pain).
13. In some hospitals, 30 to 40 percent of all deliveries are surgical.
14. In 1965, Congress established Medicare and Medicaid to cover many of the health expenses of certain cohorts of patients. Medicare pays for the health care of the elderly (those 65 years and older), the permanently disabled, and patients with end-stage renal disease. Medicaid pays for the health care of the indigent and those considered to be "medically indigent." Congress has also instituted a variety of other, less well-known programs: the Hospital Survey and Construction Act of 1946, for instance, provided funds for the construction of private hospitals in return for the hospitals' promise to provide health care services to the poor, and the Health Professions Educational Assistance Act of 1963 was created to increase the number of physicians in general, but then concentrated on increasing the number of primary care physicians and physicians who would work in under-served areas of the country.
15. The federal government has already taken a variety of steps to reduce the financial incentives to overtreat; the most notable is its Diagnostic Related Group (DRG) payment scheme.

Medical care is generally paid for in this country on a retrospective basis (i.e., the bill is calculated after services are rendered). But retrospective payment has no built-in incentives for either efficiency or cost containment. The physician can order care, necessary or not, and be fairly confident that the deep pocket of some third-party insurer will pay the bill.

In March 1983, however, Congress adopted a prospective payment scheme for handling hospital bills under Medicare (and some states also use it in their Medicaid programs). Under this system, hospitals are paid a predetermined price for their services. The amounts are arranged according to DRGs: the price is based on the average cost of treating a patient with a particular diagnosis.

It was anticipated that this new payment scheme would reduce instances of overtreatment and cut costs, and it has. Federal health spending began to moderate almost as soon as DRGs were instituted: from 1983 to 1984, Medicare expenditures grew only 8.6 percent, the smallest increase in the history of the program.

Opponents of the DRG system point out that it may have some negative side-effects as well. One concern relevant to this discussion is that instituting the DRG payment scheme will reduce access to health care: 1) DRGs may discourage hospitals from treating those people whose conditions tend to be unprofitable; 2) DRGs make it more difficult to cross-subsidize. (In the past, hospitals charged paying patients – especially those with health insurance, including Medicare coverage – more than their care actually cost. The hospitals would then use at least some of this money to cover the cost of caring for patients who could not pay for treatment. With DRGs, it becomes more difficult to shift these costs, so hospitals might be discouraged from treating indigent patients); and 3) hospitals which in the past cared for more than their fair share of the indigent and the very ill may not be able to afford to continue, and their patients may have a difficult time finding care elsewhere.

16. For a fascinating look at health care providers' use of power to secure economic advantages see Feldstein, 1988; Pauly and Redisch, 1973.
17. The Office of the Inspector General of DHHS calculated that, "Patients of referring physicians who own or invest in independent clinical laboratories received 45 percent more clinical laboratory services than all Medicare patients in general." (See Office of the Inspector General, 1989; Committee on Implications, 1986).
18. Another example of suspect decision making involves those patients receiving kidney dialysis treatments who are referred for transplantation primarily because the medical staff finds them uncooperative and wants to be free of them. (See Office of the Inspector General, 1987).

References

Anonymous. 1981. "NIH Consensus Development Task Force Statement on Caesarean Childbirth." *American Journal of Obstetrics and Gynecology* 39: 902–909.

Anonymous. 1984. "Myocardial Infarction and Mortality in the Coronary Artery Surgery Study (CASS) Randomized Trial." *New England Journal of Medicine* 310: 750–758.

Anonymous. 1988. "Forgotten Patients." *Newsweek*, August 22, 1988, p. 52.

Anonymous. 1989. "National Health Plan Wins Unlikely Backer: Business." *Wall Street Journal*, April 5.

Ansberry, C. 1988. "Despite Federal Law, Hospitals Still Reject Sick Who Can't Pay." *Wall Street Journal*, November 29.

Baily, M.A. 1988. "Economic Issues in Organ Substitution Technology." In Deborah Mathieu, ed., *Organ Substitution Technology: Ethical, Legal and Public Policy Issues*. Boulder, CO: Westview Press.

Blendon, R.J., et al. 1989. "Access to Medical Care for Black and White Americans." *Journal of the American Medical Association* 261: 278–281.

Bowler, K.M., R.T. Kudrle, and T.R. Marmor. 1977. "The Political Economy of National Health Insurance: Policy Analysis and Political Evaluation." *Journal of Health Politics, Policy and Law* 2: 100–130.

Callender, C.O., ed. 1986. "Renal Failure and Transplantation in Blacks." Transplantation Proceedings 19(supp. 2).

Committee on Implications of For-Profit Enterprise in Health Care. 1986. *For-Profit Enterprise in Health Care*. Washington, D.C.: National Academy Press.

Daniels, N. 1988. "Justice and the Dissemination of 'Big-Ticket' Technologies." In Deborah Mathieu, ed., *Organ Substitution Technology: Ethical, Legal and Public Policy Issues*. Boulder, CO: Westview Press.

Department of Health and Human Services. 1985. *Report of the Secretary*'s Task Force on Black and Minority Health. Washington, D.C.: Government Printing Office.

Duncan, R.P., and K.E. Kilpatrick. 1987. "Unresolved Hospital Charges in Florida." *Health Affairs* 6: 157–166.

Emmott, C.B., and C. Wiebe. 1989. "The Unraveling Safety Net." *Issues in Science and Technology*, 51–55.

Enthoven, A., and R. Kronick. 1989. "A Consumer-Choice Health Plan for the 1990's: Universal Health Insurance in a System Designed to Promote Quality and Economy." *New England Journal of Medicine* 320: 29–37.

Evans, R.W., et al. 1985. "The Quality of Life of Kidney and Heart Transplant Recipients." *Transplantation Proceedings* 17: 1579–1582.

Executive Summary. 1989. "Report of the National Leadership Commission on Health Care." For the Health of a Nation, Washington, D.C.: National Leadership Commission on Health Care.

Feldstein, P.J. 1988. *The Politics of Health Legislation.* Ann Arbor, Michigan: Health Administration Press.

Fisher, E.S., J.P. LoGerfo, and J.R. Daling. 1985. "Prenatal Care and Pregnancy Outcomes During the Recession: The Washington State Experience." *American Journal of Public Health* 75: 866–869.

Gleicher, N. 1984. "Cesarean Section Rates in the United States." *Journal of the American Medical Association* 252: 3273–3276, December 21.

Gortmaker, S.L. 1979. "The Effects of Prenatal Care Upon the Health of the Newborn." *American Journal of Public Health* 69: 653–660.

Gould, J.B., et al. 1989. "Socioeconomic Differences in Rates of Cesarean Section." *New England Journal of Medicine* 321: 233–239.

Greenberg, E.R., et al. 1988. "Social and Economic Factors in the Choice of Lung Cancer Treatment." *New England Journal of Medicine* 318: 612–617.

Greenspan, A.M., H.R. Kay, B.C. Berger, et al. 1988. "Incidence of Unwarranted Implantation of Permanent Cardiac Pacemakers in a Large Medical Population." *New England Journal of Medicine* 318: 158–163.

Held, P.J., M.V. Pauly, R.R. Bovbjerg, et al. 1988. "Access to Kidney Transplantation: Has the United States Eliminated Income and Racial Differences?" *Archives of Internal Medicine* 148: December.

Himmelstein, D.U., Steffie Woolhandler, and the Writing Committee of the Working Group on Program Design. 1989. "A National Health Program for the United States: A Physicians' Proposal." *New England Journal of Medicine* 320: 102–108.

Institute of Medicine, Committee to Study the Prevention of Low Birthweight. 1985. *Preventing Low Birthweight.* Washington, D.C., National Academy Press.

Joyce, K., G. Diffenbacher, J. Green, and Y. Sorokin. 1983. "Internal and External Barriers to Obtaining Prenatal Care." *Social Work and Health Care* 9: 89–96

Kahan, B.D., et al. 1987. "Complications of Cyclosporine-Prednisone Immunosuppression in 402 Renal Allograft Recipients Exclusively Followed at a Single Center for from One to Five Years." *Transplantation* 43: 197–204.

Klein, L. 1971. "Nonregistered Obstetric Patients." *American Journal of Obstetrics and Gynecology* 110: 795–802.

Ludtke, M. 1989. "Physician, Inform Thyself." *Time*, June 26, p. 71.

Moore, T.R., W. Origel, T.C. Key, and R. Resnik. 1986. "The Perinatal and Economic Impact of Prenatal Care in a Low-Socioeconomic Population." *American Journal of Obstetrics and Gynecology* 154: 29–33.

Murray, J.L. and M. Bernfield. 1988. "The Differential Effect of Prenatal Care on the Incidence of Low Birth Weight Among Blacks and Whites in a Prepaid Health Plan." *New England Journal of Medicine* 319: 1385–1391.

Nazario, S.L. 1988. "High Infant Mortality is a Persistent Blotch on Health Care in the U.S." *The Wall Street Journal*, October 19.

Office of the Inspector General, Department of Health and Human Services. 1986. *The Access of Foreign Nationals to U.S. Cadaver Organs.* Boston, MA.

Office of Inspector General, Department of Health and Human Services. 1987. *The Access of Dialysis Patients to Kidney Transplantation.* Washington, D.C.: Government Printing Office.

Office of the Inspector General, Department of Health and Human Services. 1989. *Financial Arrangements Between Physicians and Health Care Business: Report to Congress.* Washington, D.C.: Government Printing Office.

Office of Technology Assessment. 1984. *Medical Technology and the Costs of the Medicare Program, Report No. OTA-H-227.* Washington, D.C.: Government Printing Office.

Pauly, M., and M. Redisch. 1973. "The Not-For-Profit Hospital as a Physicians' Cooperative." *American Economic Review* 63: 87–99.

Placek, P.J., S. Taffel, and M. Moien. 1983. "Caesarean Section Delivery Rates: United States, 1981." *American Journal of Public Health* 73: 861–862.

Poland, M.L., J.W. Ager, and J.M. Olson. 1987. "Barriers to Receiving Adequate Prenatal Care." *American Journal of Obstetrics and Gynecology* 157: 297–303.

President's Commission for the Study of Ethical Problems in Medicine and Biomedical and Behavioral Research. 1983. *Securing Access to Health Care*. Washington, D.C., Government Printing Office.

Rahbar, F., J. Momeni, A. Fomufod, and L. Westney. 1985. "Prenatal Care and Perinatal Mortality in a Black Population." *Obstetrics and Gynecology* 65: 327–329.

Randal, J. 1982. "Coronary Artery Bypass Surgery." *Hastings Center Report* 12: 13–18.

Relman, A.S. 1989. "Universal Health Insurance: Its Time Has Come." *New England Journal of Medicine* 320: 117–118.

Sloan, F.S., J. Valvona, and R. Mullner. 1986. "Identifying the Issues: A Statistical Profile." In F. Sloan, J. Blumstein and J. Perrin, eds. *Uncompensated Hospital Care: Rights and Responsibilities*. Johns Hopkins University Press, pp. 16–53;

Starr, P. 1982. *The Social Transformation of American Medicine*. New York: Basic Books.

Starzl, T. E. 1983. "Clinical Aspects of Cyclosporine Therapy: A Summation." *Transplantation Proceedings*, Supplement 1, 15: 3103–3107.

Stone, W.J. 1983. "Medical Complications of End-Stage Renal Disease." In William J. Stone and Pauline L. Rabin, eds., *End-Stage Renal Disease*. New York: Academic Press, pp. 57–58.

Svensson, C.K. 1989. "Representation of American Blacks in Clinical Trials of New Drugs." *Journal of the American Medical Association* 261: 263–265.

Swartz, K. 1989. *The Medically Uninsured: Special Focus on Workers*. The Urban Institute Press.

Task Force on Organ Transplantation. 1986. *Organ Transplantation: Issues and Recommendations*. Washington, D.C.: Government Printing Office), p. 17.

Teffer, S., P. Placek, and T. Liss. 1987. "Trends in the U.S. Cesarian Section Rate and Reasons for the 1980–85 Rise." *American Journal of Public Health* 77: 955–959.

Weissman, J., and A.M. Epstein. 1989. "Case Mix and Resource Utilization by Uninsured Hospital Patients in the Boston Metropolitan Area." *Journal of the American Medical Association* 261: 3572–2376.

Wenneker, M.B., and A.M. Epstein. 1989. "Racial Inequalities in the Use of Procedures for Patients with Ischemic Heart Disease in Massachusetts." *Journal of the American Medical Association* 261: 253–257.

Williams, R.L., and P.M. Chen. 1983. "Controlling the Rise in Caesarean Section Rates by the Dissemination of Information From Vital Records." *American Journal of Public Health* 73: 853–867.

Chapter 6

When Enough is Enough – How to Say No to Technology

THERESA A. CULLEN

Say, for instance, your mother was dying. A new technology used in a few hospitals would possibly save her life. However, the cost of this procedure is equal to the cost of providing routine immunizations for one hundred children for one year. You are aware that there are limited revenues available for health care expenditures in your community. How and where would you choose to spend these resources?

These personal and societal questions reflect an ongoing ethical dilemma in America: how to provide adequate medical care for a population within a cost-contained system, i.e., how to meet needs and ration care. A review of specific decisions influenced by limited resources will help illustrate the difficulties inherent in this decision-making process. As we will see, these decisions must be made within a viable ethical framework.

The Indian Health Service, a federally sponsored health care system designed to provide comprehensive health care to on-reservation Native Americans, will serve to help illustrate these problems. Recently, the Indian Health Service was forced to make a bold decision about resource allocation in parts of Arizona – only life-threatening emergencies would be transferred and treated at non-Indian health service facilities. This decision reflected the increasing medical needs of the population, an inadequate budget, and recent changes in state funding for on-reservation Native Americans. "Elective" services, as well as numerous diagnostic procedures, were put "on hold" until the following fiscal year.

In a health care system such as this, health care providers are routinely forced to prioritize medical treatments. Life-threatening emergencies that are presented to the emergency room (e.g., head injury secondary to motor vehicle accident) are immediately treated and expeditiously referred to appropriate tertiary hospitals. This scenario, however, results in depletion of limited resources, as less money becomes available for routine health care. Consequently, the following kinds of decisions must be routinely made by providers within a cost-contained system. Does the patient with a routine pap smear which reveals mild dysplasia (a potential precancerous condition) receive priority for appropriate follow-up over an elderly person with symptomatic

M. Waterstone (ed.), Risk and Society: The Interaction of Science, Technology and Public Policy, 91–96.

gallstones which require surgery? Does a person who needs a heart/lung transplant to survive due to medical conditions (e.g., primary pulmonary hypertension) get priority over the need for obstetrical ultrasound to assess fetal age and anomalies?

These dilemmas reflect the focus of the American health care system on developing scientific knowledge and technology (Hiller et al., 1988). These developments have been responsible for saving countless numbers of lives (Cohn, 1988). However, this growth in medical technology has been coupled with the development of "scarce" medical resources. As a result, in the 1990's, we have limited medical resources and competing health care needs (La Puma et al., 1988). We can no longer accept technology for its own sake. A balance between new medical explorations and the provision of adequate primary care to our nation's population needs to be reached.

The United States has the most technologically-advanced and expensive health care system in the world. However, we still fail to provide access to basic medical care to all of our citizens. The number of people without private health insurance increased 47 percent between 1980–1985. The number of people without any health insurance rose by 40 percent between 1978–1986, to 37 million (Himmelstein and Woolhandler, 1989; Woolhandler and Himmelstein, 1989). As a result, real problems happen to real people. People who run out of insurance coverage are moved from one hospital to another due to their inability to pay their bills – a process called dumping (Schiff et al., 1986). Other barriers to access develop, resulting in inability to follow up abnormal lab results, inability to get regular care for chronic diseases like hypertension, and inability to obtain routine prenatal and well baby care. In effect, as is discussed below, we are daily engaged in "soft" rationing of health care based upon an individual's ability to pay (Callahan, 1988).

In response to this current situation, there is a growing movement to develop a national health care program in the United States. Fueled by health care providers who know first-hand the results of inadequate access to basic health care, many believe that we can develop a system that will provide high quality care at an acceptable cost within a national health care system (Himmelstein and Woolhandler, 1989).

However, even within this program, there will need to be limits on access to health care. New technologies will not be available to everyone. We will still need to confront the ongoing realities of limited health care resources.

Consequently, we need to recognize the need to ration health care. Rationing entails deciding which health needs should be met first. Rationing can be categorized as hard and soft. Soft rationing, as just described, refers to the casual and unsystematic manner by which some people get care and others do not. This usually happens without public debate. Hard rationing occurs when choices are openly specified, and a decision to choose one possible health good over another occurs (Callahan, 1988).

Soft rationing is often ethically unfair. It often reflects decisions made by a few and does not allow consideration of alternatives. Hard rationing, while seemingly more difficult, is in reality only more explicit. Such an explicit system would force us to debate openly how to spend our health care dollars.

There will probably never be enough of these dollars. As a result, the United States will ration resources over time. Regardless of the type of health care system our population chooses, we must utilize hard rationing in our decision making process.

However, the U.S. health care system is often driven by technological rather than social responsibilities. The costly societal custom of promoting technology without considering its social, ethical, and economic effects can no longer be tolerated (Hiller et al., 1988). We live in a medical culture that is more concerned with high-cost, life saving technologies than with care that is life enhancing. The imperative of high technology is to rescue endangered life. Major organ transplantation and intensive care units are the most striking examples of this technology (Gilledge, 1987).

In 1985, the Report of the Massachusetts Task Force on Organ Transplantation reflected the concern of one state about the public rationing of some resources. Their fundamental guidelines, reflective of the recent conclusion from the President's Commission on the Study of Ethical Problems in Medicine, centered on society's "ethical obligation to ensure equitable access to an adequate level of health care for all" (President's Commission for the Study of Ethical Problems in Medicine, 1981: 4). The Massachusetts Task Force, utilizing this premise, concluded that "access to organ transplantation means that access to something else is likely to be restricted; seeing that its costs are met means that money for something else will disappear or not be found" (Jonsen, 1985). The Task Force concretely recommended that "transplantations of livers and hearts should only be permitted if access to this technology can be made independent of the individual's availability to pay for it and if transplantation itself does not adversely affect the provision of other higher priority health care services to the public" (Jonsen, 1985).

Higher priority procedures, defined by the total number of individuals affected and the impact of the intervention on their health status, include direct services like access to prenatal and well baby care. As a principle, the Task Force stated that "if it turns out that transplantations take resources away from higher priority health care and decrease their availability to the public, these transplantations should not be performed" (Jonsen, 1985).

Two years later, the Joint Ways and Means Committee of the Oregon Legislature confronted the issue of scarce medical resources. As a result of voter imposed monetary limits to state revenues and expenditures, the legislature faced the dilemma of how to improve access to basic health care for low income residents. The Committee voted to discontinue coverage for certain organ transplantations. As a result of this decision, 1,500 people would gain health care access, while approximately 34 would be denied medically necessary transplantation and probably die.

Public response was initially slow, but then developed into lawsuits and numerous fund raising drives for individuals denied transplantation. Many state residents believed that the services being denied could be successful in selected patients. However, John Kutz Baker, M.D., Oregon Senate President, wrote, "Is the human tragedy and personal anguish of death from the lack of an organ transplant any greater than that of an infant dying in an ICU from a preventable problem brought about by lack of prenatal care?" (Welch and Larson, 1988: 122).

Massachusetts and Oregon reflect two states where the difficult issue of scarce medical resources has been openly confronted. For each state, organ transplantation became the "target" of limited resources. Why is transplantation special? Is it only the tip of the iceberg of limited medical resources (Englehardt, 1989)?

Organ transplants capture our interest for numerous reasons. They are an easily identifiable, major medical expense. They appear to benefit only a limited number of people (Grennick, 1988). Furthermore, acute rescuer health care is more valuable and subsequently more appealing to both society and individuals than preventive care (Englehardt, 1989).

Albert Jonsen refers to this life-saving compulsion on the part of American society as the rule of rescue. (See Chapter 4, this volume.)

> Our moral response to the influence of death demands that we rescue the doomed. We throw a rope to the drowning, rush into burning buildings to snatch the entrapped, dispatch teams to search for the snowbound. This rescue morality spills over into medical care, where our ropes are artificial hearts, our rush is the mobile critical care unit, our teams the transplant services. The imperative to rescue is of great moral significance, but the imperative seems to grow into a compulsion, more instinctive than rational (Jonsen, 1986: 174).

Furthermore, it is easier to target newer medical therapies than older ones for elimination. Basic medical care is usually viewed as an established good, whereas transplants are still novel and open to criticism. Basic care seems able to use public funds to supply access to the greatest number of people as reflective of the greatest good. We must acknowledge that we are responding to special pressures when we choose to provide life saving treatment to select people at the expense of providing "basic" care to a defined population (Englehardt, 1989).

In June of 1989, the *New England Journal of Medicine* published two articles that examined medical therapies designed to treat or prevent pulmonary and bowel complications in infants of very low birth weight. Both studies addressed complex medical problems in difficult to manage small neonates. Even though both studies have been criticized for faults in their experimental design, both hospital centers achieved positive results as evidenced by increased pulmonary and neurological developmental status in one study, and decreased risk of necrotizing enterocolitis (NEC – a known and potentially lethal gastrointestinal complication in preterm infants) in the other (Cummings et al., 1989; Cassady et al., 1989).

Both NEC and bronchopulmonary dysplasia (BPD – a chronic lung disease of neonates) can result in devastating medical and economic problems (Stahlman, 1989). The development of ways to treat these problems are a medical priority, as improved therapy could dramatically improve the outcome of low birthweight infants in Neonatal ICUs. However, this situation once again reflects heroic measures, and the practice of "crisis" medicine.

Prematurity is in many instances a reflection of inadequate health care for the mother over time. As described by Mathieu (in Chapter 5 of this volume), the majority of preterm infants are born to women who can easily be identified by their socioeconomic situations – women who are poor, teenaged, unemployed, undereducated, or who are involved in dysfunctional relationships. In order to prevent broncho pulmonary dysplasia or necrotizing enterocolitis, we need to prevent pregnancies which result in very low birthweight neonates.

As noted in a recent editorial

> until we set priorities and find ways to offer Americans a good education, which can lead to job opportunity, decent housing, the availability of adequate medical care, and the ability of families to live without a daily struggle for survival and without being on the welfare rolls generation after generation, prematurity will continue, with all its costly consequences. It may be only a second generation of healthy mothers who will be able to bear healthy children. We are wasting precious time (Stahlman, 1989: 1553).

We are not an equal society. The medically uninsured and underinsured do not share equally in the health and wealth of America. Most Native Americans and other minorities have only limited access to health care. Many health care providers and patients have daily knowledge of our current system, which is seemingly deaf to the voices of the indigent.

We must provide basic health care to all and at the same time be prepared to set firm upper limits to what we can provide. This would result in a health care system that is not inherently expansive. It would be a system which would balance personally limiting life expectancy to what is less than technologically possible, while providing more coherent, rounded coverage for the majority. This system could provide guaranteed access to basic, primary care for all, but would not give health an inordinately high priority in human life for a few.

References

Callahan, D. 1988. "Meeting Needs and Rationing Care." *Law, Medicine, and Health Care* 16: 261–266.

Cassady, G., et al. 1989. "A Randomized Controlled Trial of Very Early Prophylactic Ligation of the Ductus Arteriosus in Babies Who Weigh 1000 gms or Less at Birth." *New England Journal of Medicine* 320: 1511–1516.

Cohn, L. 1988. "Decades of High Tech Health Care." *Chest*, 93: 864–867.

Cummings, J.T., et al. 1989. "A Controlled Trial of Dexamethasone in Preterm Infants at High Risk for Bronchopulmonary Dysplasia." *New England Journal of Medicine* 320: 1505–1510.

Englehardt, H.T. 1989. "Shattuck Lecture – Allocating Scarce Medical Resources and the Availability of Organ Transplantation." New *England Journal of Medicine* 311: 66–71.

Gilledge, D. 1987. "Medical Technology and Some of Its Problems." *Psychiatric Medicine* 24: 257–262.

Grennick, A. 1988. "Ethical Dilemmas in Organ Donation and Transplantation." *Critical Care Medicine* 16: 1012–1018.

Hiller, et al. 1988. "Ethical Issues in Health Care Delivery." *Journal of Health Administration Education* 6: 251–261.

Himmelstein, D.U., and S. Woolhandler. 1989. "A National Health Program for the United States." *New England Journal of Medicine* 320: 102–108.

Jonsen, A. 1985. "Organ Transplants and Principle of Fairness." *Law, Medicine, and Health Care* 13: 37–40.

Jonsen, A. 1986. "Bentham in a Box: Technology Assessment and Health Care Allocation." *Law, Medicine, and Health Care* 14: 172–174.

LaPuma, J., et al. 1988. "Ethics, Economics and Endocarditis." *Archives of Internal Medicine* 1148: 1809–1811.

President 's Commission for the Study of Ethical Problems in Medicine and Biomedical Research. 1981 U.S. Government Printing Office, Washington, D.C., p. 61.

Schiff, R.L., et al. 1986. "Transfers to a Public Hospital: A Prospective Study of 467 Patients." *New England Journal of Medicine* 314: 552–557.

Stahlman, M. 1989. "Medical Complications in Premature Infants." *New England Journal of Medicine* 320: 1551–1553.

Welch, W.G., and E.B. Larson. 1988. "Dealing With Limited Resources: The Oregon Decision to Curtail Funding for Organ Transplantations." *New England Journal of Medicine* 319: 121–123.

Woolhandler, S., and D.W. Himmelstein. 1989. "Resolving the Cost/Access Conflict: The Case for a National Health Program." *Journal of General Internal Medicine* 4: 382–387.

PART III

Nuclear Power and Nuclear Waste Disposal

Nuclear power has been a subject of controversy since the beginning of the atomic age almost fifty years ago. As a society, we have been unable to achieve consensus regarding the desirability or safety of the technology, its use, or its by-products. The public image of generating electric power through nuclear fission has gone from the overly optimistic promise of "too cheap to meter" in the early days, to doom and gloom more recently.

At the beginning of the 1990's, we are seeing a tremendous resurgence of interest in nuclear power, partly motivated by predictions of global warming, partly motivated by a desire to decrease dependence on foreign (particularly, Middle-East) oil. As a consequence, we are hearing increasing numbers of comments regarding "passively" or "inherently" safe, new designs. In 1989-1990, we also began hearing guardedly optimistic reports, first from Utah and then from Texas, of initial successes with cold fusion processes.

And yet, many questions remain unanswered for the public and for policy makers. How safe is nuclear power production? How safe can it ever be? How safe does it have to be? What are the probabilities of accidents? What are the potential consequences? And finally, how will we deal with the issues of decommissioning plants and disposing of nuclear waste?

This last question raises a key ethical and moral issue: what, if anything, do we owe to future generations? As a society, we have shown a remarkable willingness to transfer risks either spatially (i.e., to other parts of the globe) or temporally (i.e., to future generations) rather than deal with them ourselves in the here and now.

Perhaps this practice has reached its most extreme manifestation with the nuclear waste issue. We have produced materials, as by-products of nuclear power generation, that must be isolated from populations for thousands of years. Even if nuclear power were eliminated immediately (a course not supported universally by any means), the problem of waste storage would still be enormous. How and where to store (since disposal is an inappropriate term) these materials poses very complex scientific and technical challenges. The institutional, equity and ethical questions are equally daunting.

Whatever decisions are finally made will have to reflect the best scientific judgments available, but will also have to incorporate an understanding of basic human values. This section seeks to illuminate several facets of this complex topic.

In Chapter 7, James Asselstine et al. explore several areas of weakness in the nuclear industry, as currently constituted. They then offer specific prescriptions for changes that would be necessary to make the industry viable. Finally, the chapter assesses the likelihood that such changes will be effected in a timely way.

Morris Farr, in Chapter 8, addresses several problems related to public perceptions of nuclear risks, and offers an educator's solution.

And finally, Ronald Milo, in Chapter 9, approaches the issues from an ethical and philosophical perspective and questions whether nuclear power is worth the risks it seems to pose.

Chapter 7

The Future of the Nuclear Power Industry in the United States

JAMES K. ASSELSTINE with SUSANNA EDEN and MARVIN WATERSTONE

Introduction

This chapter addresses the risk to society of nuclear power and nuclear waste disposal; the acceptability of those risks; the effectiveness of governmental and industry efforts to control those risks in a manner that preserves the availability of the benefits of nuclear power; and steps that may be needed to improve our management of this technology in a way that will ensure its continued availability.

More specifically, the chapter addresses the following questions, particularly as they apply to the nuclear industry in the United States:

(1) What is the risk of nuclear power and how does that risk compare with alternative means for generating electricity, or with other risks that we face?
(2) Can the risk of nuclear power be maintained at a level that is acceptably low to allow nuclear power to remain a viable energy option for the future?
(3) Have governmental and industry controls been effective in preserving nuclear power as a viable energy option for the future, and if not, what needs to be done to restore its viability as an energy option?
(4) What are the special challenges of nuclear waste disposal; how effectively have we met those challenges; and what needs to be done to improve our efforts?

Overview

There are currently 109 licensed commercial power reactors in operation in the U.S. An additional six reactors are in the latter stages of construction or are somewhere in the process of completing the Nuclear Regulatory Commission's (NRC's) operating license requirements. Thus, by about 1995 the U.S. is likely to end up with a complement of about 115 operating reactors. There are a few other nuclear units that are still listed as on-going construction projects, but actual work on them has been suspended for some time, and at this point their completion and eventual operation is doubtful.

M. Waterstone (ed.), Risk and Society: The Interaction of Science, Technology and Public Policy, 101–120.

There have been no new orders for nuclear units in the U.S. since 1978, before the Three Mile Island accident, and a substantial number of ordered nuclear units has been cancelled since that time. For a variety of reasons, new orders for nuclear plants are not expected for at least the next several years and perhaps never. The population of operating nuclear reactors in the United States by the turn of the century, therefore, will probably be limited to the present population of about 115 plants. That is a far cry from the industry and government projections made in the early 1970's that somewhere between 500 and 1,000 nuclear power plants would be operating in the U.S. by the year 2000.

Although much more modest than those projections, our current population of 115 plants still represents a sizeable commitment to nuclear power. In fact, the United States commercial nuclear power program remains the single largest in the world by a substantial margin. Together these plants generate more than 15 percent of our electricity and represent a total capital commitment of more than one hundred billion dollars (Wagner and Ketchum, 1989). In some areas of the country nuclear power accounts for 50 percent or more of the electricity consumed. Not only in those areas, but for the country as a whole, it would be extremely difficult to replace the existing nuclear plants in the short term. Substantial economic penalties and electric reliability problems would likely result.

Currently, these plants have an expected operating life of 40 years, and only a handful of the existing plants have yet reached the halfway point. In addition, the nuclear industry and the government's regulator of the industry, the NRC, have programs underway to explore the feasibility of extending the operating life of a nuclear unit by an additional 10 to 20 years. Thus, a nuclear unit entering commercial service today could still be in operation by about the middle of the next century.

In sum, even without further commitments to additional nuclear plants, we are likely to remain dependent on nuclear power for a substantial portion of our electric energy for some time. The challenge, then, is to insure that this technology is developed and controlled in a manner sufficient to preserve its availability over the coming decades. The three most relevant, and interrelated considerations in this regard are insuring the safe and reliable operation of the existing and any future plants over the coming decades; restoring the viability of nuclear power as an electric energy option and taking the steps needed to do that; and achieving a safe and environmentally acceptable solution to the nuclear waste disposal problem. To assess these issues, we now wish to address the questions enumerated at the beginning of the chapter.

Nuclear Power Risk

In general terms the risks associated with nuclear power production can be divided into two sets of concerns. The first are the risks of nuclear power under normal operations, and the second are accident-related risks.

Under normal operations the risks from nuclear power plants are relatively benign. There are a few environmental consequences, which include thermal effects of heating and cooling water and low-level radiation releases offsite from the plant. There are some effects in terms of worker occupational exposures for the approximately 90,000 workers who have now worked for the nuclear power industry. And there are also some limited fuel cycle effects from uranium mining and milling. But the big risks associated with nuclear power are not from normal operations.

The second set of nuclear power risks relate to accident risk. This is the real risk of nuclear power. In this category, some of the potential consequences include:

(1) health and safety risks; including risk to plant workers, and, in extreme cases, near-term and long-term health consequences for the public;
(2) environmental risks; which in extreme cases can result in the need to abandon land areas for several decades; and
(3) economic risks which can result in several billions of dollars in property losses onsite and, in extreme cases, offsite.

Generally, nuclear power has had an extremely successful and excellent safety record, both in the United States and worldwide, but there are two significant milestones along that path that are worth exploring because of what they can tell us about nuclear accident risk. These are the Three Mile Island accident in this country in 1979 and the Chernobyl accident in the Soviet Union in 1986. Those two accidents demonstrate the range of potential consequences that can be associated with a severe nuclear accident.

In the case of Three Mile Island, there were few, if any, off-site consequences or effects of the accident, apart perhaps from the public trauma experienced by a substantial number of people who lived around the plant in central Pennsylvania. The principal consequences of the Three Mile Island accident were economic: the loss of a plant that had only achieved one year of commercial operation, cleanup costs that have exceeded a billion dollars and are only now, more than ten years after the accident, nearing completion, and some associated costs (including public suspicions) that fell on the rest of the industry.

The Chernobyl accident resulted in much more extreme consequences. In terms of health and safety, there were risks not only to plant workers and a number of immediate fatalities, but there were also long-term health consequences, both for people who lived in the vicinity of the plant in the Soviet Union and also for people who were exposed to radiation in Western Europe. In terms of environmental consequences, Chernobyl also demonstrated the extreme case, with the need to abandon a substantial amount of land around the plant, probably for several decades, simply because the radiation levels will not permit habitation by pregnant women and small children. Finally, in terms of economic consequences, Chernobyl also demonstrated the extreme in terms of not only the substantial property losses with the destruction of the unit, but also off-site consequences largely affecting the food chain in a number of Western European countries as well as the Soviet Union.

So we have with those two accidents a good example of the range of potential consequences from nuclear accidents that are really at the heart of understanding and addressing nuclear risk. How are such risks analyzed? Understanding and evaluating nuclear accident risk is difficult due to the fact that this is essentially a probabilistic analysis. The higher probability events are relatively benign, and entail limited consequences beyond economic damage to the plant itself. However, as one moves down the probability scale, (i.e., toward more unlikely events) one encounters events with enormous potential adverse consequences.

Past and current probability analyses and actual plant operating experience provide some insights about the likelihood and potential severity of reactor accidents. Existing probabilistic risk assessment (PRA) studies show the likelihood of a severe core damage accident to be somewhere between 12 and 45 percent for a population of 100 U.S.-type reactors operating over a period of 20 years (MacKenzie, 1984: 38). Some weaknesses and uncertainties in such PRAs result from the fact that it is very difficult to account for positive and negative contributions of human behavior. There are also some differences between the assumptions used in the studies and the actual performance at the plants.

The risks of off-site consequences to the public and to property depend largely upon containment performance. There is a range of performance for U.S. reactor containments with some having a very low likelihood of failure and some having a high failure risk during a severe accident. There are also severe accident sequences (some discovered only recently) which occur in situations where containment protection is reduced or unavailable.

Comparing Nuclear and Non-Nuclear Energy Alternatives

Such comparisons are difficult because one is comparing two different types of risk. In the case of non-nuclear energy alternatives, such as coal, oil, or natural gas the principal health and environmental risks are due to normal operations. There is evidence of safety risks from the fuel cycles (mining and transportation) for these alternatives, and growing evidence of health and environmental risks from plant emissions, particularly as our knowledge of the effects of acid rain and global warming trends increases. The 1990 Valdez accident also gives us a clear example of the potential environmental costs associated with oil transportation. Thus, under normal operating conditions, the evidence appears clear that nuclear power has health and environmental advantages over competing electric energy alternatives.

The comparison becomes more difficult when one factors in nuclear accident risk which includes the potential for high consequence but low probability events. Many in government and within the industry argue that even taking into account the potential for an accident, the overall risks associated with nuclear power are no greater than, and may well be lower than those associated with other competing energy alternatives or with floods, dam failures, and earthquakes. There is some merit to this argument. Nevertheless, one must recognize that the public views nuclear accident risk somewhat differently, and that the public is not likely to tolerate a severe nuclear accident with significant off-site consequences, even if those consequences are comparable to, or even somewhat less than, those of other natural or human-caused disasters.

Nuclear Power and Acceptable Risk

This leads to our second question, which is: what is the level of nuclear accident risk that the public is likely to find acceptable and tolerable in order to permit this energy alternative to remain viable? Our sense is the level of risk that would be acceptable is considerably lower than the current level of risk of nuclear power worldwide.

Fairly substantial changes have taken place in the nuclear power industry since the Three Mile Island accident. Modifications to the plants and the manner in which those plants are operated have cost on the order of fifty million dollars per plant over the last decade. The post-TMI changes included substantial efforts to add personnel, to upgrade their qualifications, and particularly their engineering knowledge and ability to understand what is happening in the plant under accident conditions. Personnel now are much better trained than they were before TMI (Thompson and Lippman, 1990). Plant equipment, particularly instrumentation to assist operators in understanding what 's taking place in the plant has also improved significantly, and emergency planning requirements have been put in place to provide an added measure of protection for the public (Okrent and Moeller, 1981).

The NRC and the industry also initiated severe accident reviews that were intended to examine events – precursor or warning events to future severe accidents – identify problems and get them corrected. The NRC has recently ordered individual, plant-by-plant evaluations to review, identify and correct severe accident vulnerabilities at each individual plant. Because each plant in the U.S. was essentially designed from scratch, there are different vulnerabilities for each plant, even in cases in which the reactor manufacturer was the same. There also has been considerably more attention paid since Three Mile Island to learning the lessons of operating experience.

The NRC and the industry have also focused attention on containment performance. There is a wide range of expected performance among reactor containments. Those plants that have large dry containments have a high likelihood, at this point in our understanding, of retaining their integrity even under severe accident conditions. At the same time, some of the smaller, weaker containments, including those of the GE boiling water reactors with the Mark

I containment design and the Westinghouse ice condenser plants, have a much higher probability of failure under severe accident conditions.

Taking all those things together, the trend is toward improved safety performance at the plants. At the same time, the probabilistic risk assessments that have been done, the precursor events that we have identified over the past decade, and actual plant operating performance still tend to indicate that there is a substantial likelihood of a severe accident at a U.S. plant over the next twenty or more years, and certainly over the remaining operating life of the plants that we now have (MacKenzie, 1984).

In terms of what can be done to further reduce that risk, it is possible to build on the steps that have been taken over the past decade. With a few additional efforts we could substantially reduce that risk and help to ensure that we do not see a severe accident that has any off-site consequences for the public.

First, we need somewhat more formal programs to address such areas of weakness as personnel training, equipment reliability and maintenance. These still tend to be somewhat weak throughout the industry. We also need detailed knowledge of accident vulnerabilities at each individual plant so that these can be corrected.

Second, we could give some further consideration to severe accident prevention and to mitigation measures that are now actively being considered or adopted by a number of European countries. On the accident prevention side, a number of European countries, including Sweden, Germany, and Switzerland, have already adopted or are adding to some of their older plants, additional decay heat removal systems and additional emergency electrical supply systems. These modifications are intended to help improve the reliability and degree of assurance that under an accident condition it will be possible to continue to provide cooling water to the core and have the electricity to operate those cooling systems. In terms of accident mitigation measures, again countries such as Sweden, Switzerland, Germany, and France are either considering or are already installing filtered, vented containment systems to provide an added level of assurance that under severe accident conditions there would not be large, harmful off-site releases of radiation (International Atomic Energy Agency, 1988; Becker, 1989).

With those kinds of measures, building upon what has occurred over the past decade or so, nuclear power in the U.S.could also be brought to the point where it would be possible to ensure a very, very low likelihood of a severe accident over the remaining operating lives of the plants that we now have and an extremely low likelihood of a large off-site release in the event of a severe accident. That must be the standard that we continue to strive to achieve, if nuclear power is to be acceptable as an alternative to other energy technologies.

The Role of Government and Industry

This brings us to the third question: have governmental and industry controls been effective in preserving nuclear power as a viable energy option for the future, and if not, what needs to be done to restore its viability? Our argument is that they have not. It is clear that at some point utilities in the U.S. will need additional electric generating capacity, and it is likely that this need will eventually include new central station, base-load generating facilities. Whether nuclear power will be an available option at that time will depend upon the confidence that the utilities, their shareholders and investors, state rate regulators, and the public have in nuclear power. In turn, confidence depends upon safety, technical, and financial factors, as well as confidence in our ability to dispose of nuclear wastes in a safe and environmentally acceptable manner.

At present, we would argue that the necessary confidence is lacking in all quarters. Public concerns focus on safety, the nuclear waste disposal problem, and the high cost of some new nuclear plants in this country. On the safety side, since the Chernobyl accident in 1986, public opinion polls have indicated substantial public opposition to the construction of new nuclear plants in this country, although the public attitude toward existing plants is more supportive, or at least more ambiguous (Hohenemser and Renn, 1988; Wagner and Ketchum, 1989; Wald, 1990).

As for the nuclear waste problem, public concerns take two forms. One concern is that our continued inability to develop a long-term storage or disposal solution to the waste problem can turn existing and future nuclear plants into de facto spent fuel storage sites. A second concern on the part of people in the states under consideration for a waste storage or disposal facility is that such a facility may pose safety or environmental risks or be harmful to the local economy. We will return to these issues in the final section of the chapter.

Utility, shareholder, investor and rate regulator concerns focus on the unpredictability of the process for designing, building, licensing, operating, and ultimately, decommissioning a new nuclear power plant in this country.

Additionally, utilities with existing nuclear plants are suffering from what can perhaps best be described as nuclear burn-out. These utilities are finding that building, licensing, and operating nuclear plants poses greater challenges, and requires greater resources and much more management attention than is the case for competing alternatives. As one senior utility executive of a company with a very successful record of nuclear plant construction and operating performance said a few years ago, "I 'm sick to death of the problems of nuclear power." This experience with the existing plants obviously affects a utility 's willingness to consider a future commitment to new plants.

If confidence on the part of the utilities, their shareholders and investors, rate regulators, and the public does not support a commitment to new nuclear plants at this time, what needs to be done to restore that confidence? There are seven factors that must be addressed to restore this confidence. These factors are: 1) the safety and reliability performance of the 115 plants now in operation or in the latter stages of construction in this country; 2) the economic recovery by the utilities of their investment in the existing plants; 3) the development of a better product for the future; 4) the development of a better regulatory process; 5) the development of a better process for building the plants and for overseeing plant construction; 6)greater risk sharing in the financing of new nuclear units; and finally, 7) a special concern, the nuclear waste disposal problem. The rest of the chapter addresses these factors.

Safety and Reliability Performance of the Existing Plants

This is probably the single most important element in determining whether additional nuclear plants will be built in the United States (and to some extent overseas). There are really two components to this discussion, the first being the risk of an accident, and the second being more routine plant operating performance. Clearly, the first and most significant objective regarding the safety performance of the plants is to ensure an accident-free operating record for the plants for the next decade and beyond. Our experience with the Three Mile Island accident demonstrated that plant accidents can have disruptive and costly consequences for the entire industry.

A study by the Atomic Industrial Forum (1984), the nuclear industry's trade association assessed (in the aftermath of the TMI accident) construction, and operating and maintenance activities for existing nuclear power plants during the period from 1980 to 1984 and identified rising costs for these activities. Much of this increase in costs can be attributed to the need to address the lessons and new requirements generated by the accident. For new plants under construction, the post-TMI years were a period of turmoil and uncertainty. The post-TMI changes were disruptive of the construction program and increased costs for each of these plants.

It is reasonable to expect much the same reaction in the event of another nuclear plant accident in this country. In an effort to address public concerns and to justify the continued operation of the remaining plants, the regulators can be expected to impose a series of new safety requirements which inevitably must be developed within a short time period and under crisis conditions. Such requirements will not necessarily be well thought-out or carefully scrutinized. Demands will be made on the utilities to meet these new requirements on an expedited basis. The result will likely be a period of further disruption and uncertainty, with major new expenses for the industry. Such an environment will discourage any utility interest in new nuclear plant investments.

The future viability of the nuclear option is also likely to be affected by the operating performance of the existing plants over the next several years. Put simply, if the utilities are unable to achieve good safety and reliability performance from their existing nuclear plants, they are unlikely to order new ones. An examination of the operating performance of our nuclear power plants over the past several years demonstrates that there is a broad range of performance among U.S. utilities (Gordon et al., 1987; Boley et al., 1988). Some plants have established excellent safety and reliability performance records and rank among the top performers worldwide. Alabama Power's Farley units, Northern States Power's Monticello and Prairie Island units, Wisconsin Electric's Kewaunee plant, Wisconsin Public Service's Point Beach units, Northeast Utilities' Millstone plant, Duke Power's Oconee units, Florida Power and Light's St. Lucie units, and Yankee Atomic's Yankee-Rowe plant are examples of U.S. nuclear plants with consistently good operating and safety performance.

But there is another group of plants that falls at the other end of the performance spectrum. This group includes TVA's Browns Ferry and Sequoyah plants, Sacramento Municipal Utility District's Rancho Seco plant, Boston Edison's Pilgrim unit, Philadelphia Electric's Peach Bottom Units, Florida Power and Light's Turkey Point units, Colorado Public Service's Ft. St. Vrain plant, Detroit Edison's Fermi 2 unit. Improving the performance of the plants which fall on the low end of the scale and assuring a much greater degree of uniformity and consistency in nuclear operating performance is one of the principal challenges facing the industry and the NRC today. Although there are signs of progress in the case of several of the plants mentioned, there is still considerable room for improvement.

A review of operating experience over the past few years also indicates that we are experiencing on the order of two or three significant operating events per year. These events frequently lead to the extended shutdown of the plant both to correct the problems which cause the event and to restore the NRC's confidence in the utility's ability to operate the plant safely. In addition, the NRC has begun to identify plants with a history of poor operating performance. Of these, about half have experienced extended forced shutdowns.

The recent record of operating experience with the plants also discloses opportunities for general improvement throughout the industry in several areas. These include personnel performance, equipment reliability, and maintenance. Finally, two other phenomena are emerging, particularly at the older plants, which can affect operating performance. They are the appearance of design problems, either dating from the original licensing of the plant or resulting from the plant modification process over the years, and the effects of plant aging. Examples of the latter include deterioration in plant piping systems, degradation in steam generators, and deterioration of electrical equipment in plants which have completed only one-quarter to one-third of their expected operating lives. How well the industry and the NRC address these existing and emerging operations issues will have a significant bearing on the future operating performance of the plants (Golay, 1984; Lester, 1984; Lidsky, 1984).

It is also becoming increasingly clear that there is a direct link between safety performance and reliability performance. In general, the plants with poor histories of safety performance also have poor reliability records. In addition, the poor safety performance and extended shutdowns experienced by some plants have contributed to the generally lower overall U.S. nuclear plant capacity factors. The economic consequences of poor safety and reliability performance are also becoming increasingly apparent. Performance, and the potential adverse economic impacts of poor plant performance, are factors that should be considered by investors in evaluating the credit quality of individual utilities, and we have incorporated this aspect of utility performance in our reviews.

What can be done to improve nuclear plant safety and reliability performance and to assure greater consistency throughout the industry? First, we need to attack the areas of weakness in current performance. This will require new initiatives aimed at improving human performance, equipment reliability, and maintenance. Although the industry has efforts underway in each of these areas, a somewhat more formal and aggressive approach is needed. Second, we need a more formal program to identify design weaknesses early, before they result in serious operating events or accidents. Third, we need to examine options for further reducing accident risk. These options should include both accident prevention and accident mitigation measures. Fourth, we need to reduce inconsistency by focusing attention on the poor performers. There are recent indications that the NRC is moving more in this direction. These indications include the Commission's criticism of Philadelphia Electric's management for the operating problems at the company's Peach Bottom plant, and the NRC's negative comments on the operating problems at Florida Power and Light's Turkey Point units.

Finally, we need to consider other options for improving the operational and technical capabilities of the utilities. One alternative worth exploring is the establishment of operating entities to assume the responsibility for operating several nuclear plants. Today, of the 53 licensees with operating nuclear plants, only 13 have responsibility for the operation of more than two nuclear units. Thus, the vast majority of nuclear utilities operate only one or two reactors. This large number of single-unit and two-unit nuclear operators adds to the lack of uniformity of operating performance within the industry and makes it more difficult for the utilities to share knowledge and learn the lessons of each other's operating experience. It also makes it more difficult for the utilities to develop and agree to industry-wide improvement initiatives, and adds to the complexity of regulation.

Moving to fewer, large operators would ease some of these difficulties and could result in greater operating efficiencies. Some utilities have begun to see the advantages in this approach. The recent merger of Toledo Edison and Cleveland Electric into Centerior Energy included the consolidation of the nuclear organizations of both utilities into a single unit with responsibility for operating the Davis-Besse and Perry reactors. This represented at least a first small step toward consolidating nuclear operations.

In addition, the Southern Company is establishing a single nuclear operating organization to operate Alabama Power's two Farley units and Georgia Power's two Hatch and two Vogtle units. The Southern Company example offers the opportunity for a broader sharing of the excellent operating strengths of Alabama Power's Farley nuclear organization.

Initiatives of this type could do much to help ensure a successful safety and reliability performance record for the industry over the next decade. Having fewer, large nuclear operating organizations which focus exclusively on the challenges of improving the safety and reliability performance of the existing plants could also help to address the utility burnout problem. If coupled with the efforts outlined to address the root causes of today's operating problems, these organizational changes could help to ease the burdens on the utilities now associated with operating nuclear units.

Utility Recovery of their Investments in the Existing Plants

We now want to turn to our second factor: the economic recovery by the utilities of their investments in the existing nuclear plants. This is a particular concern for the new, large nuclear units which have gone into operation in the past several years and for the few remaining plants in the latter stages of construction in the U.S. Utilities are unlikely to consider ordering new nuclear power plants if they cannot make the existing plants operate reliably or if they end up losing substantial portions of their capital investments in those plants. At the same time, state economic regulators are not likely to look kindly at new nuclear plants if the existing plants are not competitive generators of electricity.

We have already described the steps that would help bring about improved safety and reliability performance of the existing plants. Beyond that, what is needed is a mechanism to provide a fair return to the utilities for their prudent investments in the existing plants that does not result in excessive rate increases which threaten the health of state and local economies. This is particularly true for the large, new nuclear units which entered commercial operation in the mid-to-late 1980s. Under traditional ratemaking, the high cost of these units creates the potential for large initial rate increases which can adversely affect the state and local economy, drive away a utility's industrial and municipal customers, and substantially increase the cost of service to residential and commercial customers.

There does not appear to be a magic solution to this problem. However, there does seem to be increasing popularity, at least in some areas of the U.S., of sale and leaseback arrangements for new nuclear plants, as well as for other generating and transmission facilities. For the newer nuclear units in particular, these arrangements offer the advantage of reducing rate shock by levelizing electric rates over the operating life of the plant. This approach may make it easier for utilities to obtain favorable economic regulatory treatment for their investments in large, new nuclear units.

In addition to sale and leaseback arrangements, some utilities are exploring changes in the traditional structure of the utility industry. Some of these structural changes offer the opportunity to reduce the high initial rate treatment of new generating units. Although these new arrangements have their drawbacks as well as their advantages, they offer the opportunity for utilities to recover much, if not all, of their capital investments in the current generation of nuclear plants. That is essential if utilities are to consider future investments in nuclear units.

A Better Product for the Future

We now want to turn to our third factor. Put simply, we need a better product in the future which satisfactorily addresses the main problems we have experienced with the existing plants. Those problems include the following. First, the existing plants were designed and built without the benefit of substantial operating experience with similar plants. This is particularly true for the newer, larger generating units that have entered commercial operation over the past decade. Due to the rapid scale-up in the size of nuclear units, these larger plants were designed and built with the benefit of only limited operating experience from the early, small plants in this country, and with very little operating experience from intermediate size and larger plants. As a result, a number of safety problems were identified after construction of the plants began. In many instances, resolution of the problem was deferred until after the plants were in operation. Of course, problems that require plant modifications are much more difficult to address after the plants are physically constructed. Some of these problems are still outstanding. Unresolved safety issues such as decay heat removal, and systems interaction are examples of outstanding issues that have yet to be fully addressed for the 109 nuclear units now in operation.

Second, we paid insufficient attention to the balance of plant. High quality nuclear standards for plant design and construction were limited to the nuclear reactor itself and to some of the primary safety systems. The balance of the plant was essentially unregulated by the NRC. Consequently, many plant systems and equipment, including much equipment that can directly affect safety and reliability performance, were only designed and built to meet normal industrial standards. Those standards in some instances have proven inadequate. A high percentage of significant operating events in recent years either originate with or are complicated by poor performance with balance of plant equipment (Golay, 1984; Lester, 1984). This difference in performance between the higher quality and more carefully regulated nuclear portion of the plant and the lower quality and unregulated balance of plant systems is one of the principal contributors of today's nuclear plant performance problems.

Third, with a few limited exceptions, we failed to standardize nuclear plant design in this country (Okrent and Moeller, 1981). Our operating nuclear plants were designed by different reactor manufacturers, and numerous, different architect-engineering firms. In addition, the plant designs were tailored to accommodate the different wishes and needs of more than 50 utility operators. As a result, even plants with nuclear steam supply systems designed by the same reactor manufacturer are different, with their own individual problems and vulnerabilities. These differences in plant design substantially complicate the utilities' efforts to achieve safe and reliable plant operation.

Due to differences in plant design, equipment, and performance, it is also more difficult for the industry and the NRC to learn the lessons of experience. If problems occur at one plant, the situation at every other plant must be examined to determine whether the problem applies in that situation as well. Often, solutions must be tailored to fit individual plant circumstances, making it more difficult to develop generic solutions to safety problems.

The result is more time-consuming and often more costly than is the case with foreign nuclear programs such as the one in France, which are heavily standardized (Fagnani and Moatti, 1984; Becker, 1989). In addition, the utilities are deprived of economies that would be available through standardization, such as the sharing and ready availability of spare parts. Finally, the practice of custom designing each of the plants has added to the cost of our nuclear plants, since design costs cannot be spread over several units.

A fourth problem is that we began the construction of our existing plants with only a limited portion – typically, 12 to 20 percent – of the design work for the plant complete (Okrent and Moeller, 1981). This meant that the plants were being designed as they were being built. This design-as-you-build approach enormously complicated the construction process for the plants, often leading to poor coordination between the design and construction teams, to the need for retrofit, and to delays in plant construction. In addition, the limited amount of design information available prior to the start of plant construction encouraged the deferral of significant safety issues until later in the construction process or worse, until after the plant went into operation when those issues were much more difficult and costly to address.

The problems of the past have added significantly to the complexity and cost of the existing plants. They have resulted in a process of almost continual change and disruption, and they have created new difficulties for those utilities that have not properly controlled the plant modification process.

What can be done to ensure a better product for the future that takes advantage of these lessons of the past? First, we must standardize. The designs for any future plants should be drawn from a small number of pre-approved standardized designs. Such an approach can help to concentrate and focus the efforts of both the industry and the regulators to ensure a higher quality design.

In addition, both the industry and the NRC must make a sufficient investment of time and resources in the original design preparation and review process to identify all significant safety issues and to address them in a manner that will lead to good safety and operational performance and long-term stability. Put another way, a new standardized design should not be approved for use until all known safety issues are addressed in a satisfactory manner. Such an approach should result in the correction of problems when they are easiest to address – when the plant design is on paper but not in steel and concrete – and should reduce the potential for new and unforeseen problems in the future.

Designs for future plants should also make those plants easier to operate and maintain, and more forgiving in their safety performance. This may require that future designs be for simpler, smaller plants which are designed specifically with long-term maintenance in mind. Future designs should also emphasize quality throughout the plant to avoid a repetition of the current "balance of plant" problems.

Apart from their safety benefits, smaller plant designs may also offer the utilities other advantages. For example, smaller plant designs offer the utilities greater flexibility in adding future generating capacity in smaller increments. These may be easier for utilities to absorb than current generation 1,000 MWe plants, and may avoid the need for complicated joint ownership arrangements similar to those that we have seen for some of the large existing plants.

Finally, we should insist upon an essentially complete design before the design is approved for use in constructing a new plant. This should put an end to the added challenges, complexities and uncertainties of the design-as-you-build approach.

There are some signs for optimism that a better product will be available for the future. Both Westinghouse and General Electric are working on advanced light water reactor designs which might be ordered in the mid-1990s (Weinberg et al., 1985). The Department of Energy has also launched an effort to develop longer-term advanced reactor designs. These designs would be for much smaller, modular plants with simple plant designs and very high margins of safety (Lidsky, 1984).

Despite these positive signs, there remain substantial impediments to developing the types of new plant designs which are necessary to any future commitment to nuclear power in the U.S. Development of new, improved, and more complete standard designs along the lines that have been described will require an initial investment by the reactor manufacturers of at least several hundred million dollars (Golay, 1984; Lidsky, 1984). It is not clear that the senior managements of these companies are willing to make this added investment without the prospect of firm orders for the new product. Yet, development of an improved design is likely to be an essential precondition to any utility willingness to consider ordering a new nuclear unit.

In addition, the industry efforts thus far to set new standards and criteria for future designs indicate that there is still resistance in some quarters to making the sacrifices in plant efficiency and size that may be needed to obtain significant improvements in simplicity and plant safety margins. And finally, the development of new and improved, complete standard designs will require a high degree of coordination and cooperation among the various elements of the industry, including the reactor manufacturers, architect-engineers, constructors, and utilities. It is also not yet clear that these various entities are willing to make the sacrifices in autonomy, and cost and profit sharing needed to mount a successful design development effort.

A Better Regulatory Process

Our fourth factor is the need for an improved regulatory process for licensing new plants and for overseeing the operation of existing plants. In the past, we have used a reactive approach to regulation. Under this approach, we often failed to recognize or fully address significant safety issues. We allowed plants to be constructed and even to go into operation with a substantial number of unresolved safety issues. All too often, we waited to address safety problems until after those problems contributed to accidents or serious operating events at the plants.

This approach has had adverse consequences for the industry, for the NRC, and for the public. For the industry, this approach has contributed to a lack of stability and predictability, making regulatory standards a constantly moving target. For the public the reactive approach to regulation has raised questions about the effectiveness and objectivity of the NRC. All too often, the public has come to see the regulatory process as closed and unresponsive to their concerns or views. For the NRC, this public perception of the regulatory program has eroded the agency's image as a fair, objective, and competent regulator of the industry.

Several changes are needed if we are to have a more stable and predictable regulatory process for the future. Regulators must do a more thorough job in their initial reviews of license applications than has been the case in the past. In particular, the NRC must address all known safety issues as part of the initial license review before the start of plant construction. Moreover, these issues must be addressed in a manner that is likely to lead to long-term stability. This may require doing more than the bare minimum and building into the plant design sufficient safety margins to accommodate the resolution of new issues which might arise in the future.

In addition, NRC's license application review process should invite public observation and involvement. This would allow the public to see firsthand the give and take that occurs between the NRC staff and the utility applicant during the NRC's safety review. And it would provide an opportunity for the NRC and the utility to hear and respond to issues of interest to the public at an early stage in the process.

There is also room for improvement in the NRC licensing hearing process. As with the NRC staff's design review, attention should be focused on the early resolution of issues and on avoiding reopening issues once they have been decided. Members of the public should be given a fair opportunity to raise legitimate safety, environmental, and security issues at the earliest point in the process at which the issue can be raised. Once an issue is raised and decided, however, the issue should not be reopened absent a compelling showing that the earlier decision was incorrect. This approach would be aimed at resolving all design issues before the start of plant construction. The same would be true for all questions of site suitability and the feasibility of emergency planning. The only remaining open issues to be resolved prior to plant operation would be the adequacy of plant construction, the adequacy of the actual emergency planning measures for the plant, and the utility's capability to operate the plant safely.

As a final matter regarding the licensing process, We should require as a condition to issuing any new construction permit for a nuclear plant that the state in which the plant is to be located certify to the NRC that there is a need for the power to be produced by the plant, that the state supports the proposed nuclear plant as the preferred alternative for meeting that power need, and that the state believes that adequate emergency planning measures can be provided for the site. Although these state certifications would not be an absolute guarantee against changes in the state's position, they would represent at least a step toward some greater degree of predictability and stability in the process. And if the state has fundamental objections to the plant or its location, the time to learn of those objections is before construction work on the plant has begun rather than after the plant has been built.

A Better Process for Building Nuclear Plants

Our fifth factor is the need for a better process for building nuclear power plants and for overseeing plant construction. In the past, we have seen mixed performance in the construction management of nuclear projects in this country. Some utilities and their contractors have done an excellent job in obtaining effective and efficient management of construction of their nuclear plants. Florida Power and Light's St. Lucie 2 unit and Commonwealth Edison's four standardized Westinghouse units and Byron and Braidwood have earned reputations for good construction management and reasonable cost plants. Among utilities without prior nuclear experience, Union Electric's Callaway plant and Pennsylvania Power and Light's Susquehanna reactors come to mind as examples of effective and efficient construction management and successful plant startup.

However, some other utilities have experienced considerable difficulties in meeting the construction management challenge, and a few have faced the cancellation of plants due to construction quality assurance problems. All too often in these cases, the scope and extent of the quality problems have not been identified until the eleventh hour after construction work on the plant has been substantially completed. At that point, it is a very difficult, time-consuming, and costly process to verify that plant construction has been adequate. In such cases, complex reinspection programs, extensive rework of the plant and seemingly interminable licensing hearings are the norm.

Several changes to the existing construction management process are needed if we are to ensure consistently excellent performance for future nuclear plants. We need to pay greater attention to the qualifications and experience of the construction management team. Before construction begins, the utility and the NRC should ensure that a well-qualified and experienced construction management team is in place.

Once construction begins, the NRC and the utility should employ a step-by-step monitoring process to verify that the management team is functioning effectively. This process should consist of the review and sign-off of various activities before proceeding to the next stage of construction work. This approach builds upon the "readiness review" concept that Georgia Power has used effectively in the construction of its Vogtle units. A benefit of this approach is greater regulatory stability because NRC sign-off on the adequacy of work is provided on a continuing basis as work proceeds through the various stages of plant construction. If quality problems are identified by this monitoring approach, construction work should stop until the difficulties are corrected. Under no circumstances should quality problems be allowed to persist for years until the plant is essentially completed.

Finally, we should seek to develop a more effective quality assurance program which provides greater guarantees that construction work is being performed properly. This probably requires more emphasis on direct verification of actual work and less emphasis on paperwork than has been the case in the past. Ideally, this new quality assurance system should take into account the needs of efficient construction management with less stopping and waiting and less rework.

Greater Financial Risk-Sharing

Our sixth factor is the need for greater risk-sharing in the financing of new nuclear power plants. There is a growing perception, particularly in the case of those plants that have experienced difficulty in the construction and licensing process, that too large a share of the financial risk of the project has fallen on the utility, its ratepayers, shareholders, and creditors. If unforeseen problems arise, if the project is substantially delayed, or even cancelled, the associated financial burden tends to fall on one or more of these groups. Even after a plant has entered commercial operation, the costs associated with poor operating performance and unanticipated or deferred problems tend to fall on these groups.

For the future, if nuclear plants are to be viewed as prudent business investments, it will be necessary not only to minimize the potential for events that can involve substantial financial penalties, but also to provide some mechanism for greater risk- sharing of those penalties by other elements of the nuclear industry if such events occur. At least two options for greater risk-sharing are arrangements which involve nuclear suppliers as co-owners or partners in the project and stronger warranties of performance for their products. The recent announcement by Consumer Power of the proposed transfer of the utility 's Palisades plant to a joint venture involving the utility and Bechtel is an example of a joint ownership arrangement for an existing plant (Lippman, 1990). This type of approach for a new nuclear unit would give the nuclear supplier a much more direct financial stake in the success of the project and would reduce somewhat the financial exposure of the utility and its shareholders, creditors, and ratepayers.

In terms of warranties, some foreign nuclear suppliers already provide much more extensive warranties than is the case in this country. For example, ASEA-Atom, Sweden's nuclear manufacturer, warrants that it will complete its nuclear units on a specified schedule and at a specified cost, and that the plants will meet predetermined levels of safety and reliability performance. Better than expected performance is rewarded by incentive payments to the manufacturer and poorer performance results in financial penalties.

Warranties of this type and other mechanisms for future nuclear plants in this country would provide another means of financial risk-sharing and would involve the utility and its suppliers in a more direct partnership to bring about successful and timely plant construction, licensing, and operation (Weinberg et al., 1985). They would also be a positive step toward convincing a utility's directors, shareholders, creditors, and rate regulators that the benefits to be gained from a new nuclear unit are worth the potential financial risk.

The Nuclear Waste Disposal Problem

We want to close with a few points on an area of special concern: the problem of nuclear waste disposal. Our continuing inability to make significant progress toward the development of a safe and environmentally acceptable nuclear waste disposal facility affects the future viability of nuclear power in two ways. First, it threatens to erode public confidence in nuclear power even below current levels. This threat comes both from strong local opposition to a repository in the states under consideration as possible hosts and from a growing perception that our inability to develop a repository may turn individual reactor sites into long-term spent fuel storage sites. Our inability to solve the waste problem threatens the continued operation of the existing nuclear plants in this country and the licenseability of any future plants. The courts have required, both as a condition to issuing new plant operating licenses, and in order to allow the continued operation of the existing plants, that the NRC determine it has confidence that a solution to the nuclear waste storage and disposal problem will be available when needed.

Thus far, the NRC has been able to make this "waste confidence" finding based in large measure on the enactment of the Nuclear Waste Policy Act of 1982 (PL 97–425). This finding must be reviewed on a periodic basis, and the current difficulties in implementing the 1982 Act could adversely affect the NRC's ability to make a positive finding in the future. Growing public concern about the waste problem, opposition to the development of the needed disposal facilities, and uncertainties about the continued operation of existing plants and the licenseability of future plants all serve as a substantial deterrent to any future commitments to nuclear power in this country.

The principal difficulty with the Department of Energy's program for solving the nuclear waste disposal problem is that DOE's program lacks credibility with the states being considered as locations for a repository. This lack of credibility extends to both the technical adequacy of the program and the fairness of the process. Among other things, the states argue that DOE is not following a careful and conservative technical approach and that technical decisions are driven by a commitment to meet an arbitrary schedule. In addition, the states argue that DOE's siting decisions are motivated by political considerations and not by technical merit. As a result of these difficulties, the schedule for developing a repository has slipped by at least one year for each of the years since the Nuclear Waste Policy Act was passed in 1982 (see GAO, 1987a; 1988).

The growing state opposition and the failure to make progress toward achieving a successful result mean that it is time to review the program and get it back on track. This effort ought to be directed toward restoring technical merit as the basis for decisionmaking in the repository program, restoring the integrity of the site selection process, and ensuring that a credible and effective organization is managing the program. In the summer of 1987, Congressman Morris Udall and others introduced legislation which would provide for such a review of the program. This legislation is a positive step toward resolving the current impasse.

Although there were competing proposals being considered in 1987, the Congress decided to proceed with site development for a single site for the first repository; a decision that was ill-advised. Relying on a single site creates the potential for failure if unanticipated problems come to light during the site exploration and examination effort. Our unfortunate past experience with the Lyons, Kansas site demonstrates the danger in focusing our efforts on only one site, and we can ill-afford a repetition of that failure (GAO, 1987b).

Conclusion

We have attempted to provide a list of the conditions that must be met if nuclear power is to once again be a viable option for meeting future energy needs in the U.S. That list is a formidable one, and each element poses some rather significant challenges. Satisfying these conditions will not be easy and will require a concerted effort by all sectors of the industry as well as some rather fundamental changes in the manner in which we have gone about designing, building, financing, and operating nuclear plants in the U.S. It will also require a new approach to safety regulation and plant licensing – one that is more forward looking and effective, and that emphasizes long-term stability and predictability. And finally, it will require significant progress toward a solution to the nuclear waste problem. Each of these tasks is achievable. Our experience over the next few years will tell whether we are up to the challenge and whether our efforts are likely to be successful.

References

Atomic Industrial Forum. 1984. *Positive Experiences in Constructing and Operating Nuclear Power Plants Worldwide*. Washington, D.C.: Atomic Industrial Forum.

Becker, K. 1989. "International Harmonization of Nuclear and Radiation Standards and the Future of Nuclear Energy." *Energy Sources* 11: 85–94.

Boley, K., et al. 1988. *Who's* at the Controls? A Critique of the Training and Qualifications of Nuclear Power Plant Operators. Public Citizens' Critical Mass Energy Project Report.

Fagnani, J., and J.P. Moatti. 1984. "The Politics of French Nuclear Development." *Journal of Policy Analysis and Management* 3 (2): 264–275.

GAO (General Accounting Office). 1987a. *Nuclear Waste: Institutional Relations Under the Nuclear Waste Policy Act of 1982*. GAO/RCED–87–14. Washington, D.C.: GAO.

GAO (General Accounting Office). 1987b. *Nuclear Waste: Status of DOE's* Implementation of the Nuclear Waste Policy Act. GAO/RCED–87–17. Washington, D.C.: GAO.

GAO (General Accounting Office). 1988. *Nuclear Waste: Fourth Annual Report on DOE's* Nuclear Waste Program. GAO/RCED–88–131. Washington, D.C.: GAO.

Golay, M.W. 1984. "An Agenda for Improving Present-Day Reactors." *Technology Review*, February/March: 49–51.

Gordon, J., et al. 1987. *Nuclear Power Safety Report*. Public Citizens' Critical Mass Energy Project Report.

Hohenemser, C. and O. Renn. 1988. "Chernobyl's Other Legacy." *Environment* 30 (3): 4–11, 40–45.

International Atomic Energy Agency. 1988. *Nuclear Power and Fuel Cycle: Status and Trends*. Division of Nuclear Power and Division of Nuclear Fuel Cycle. Vienna: IAEA.

Lester, R.K. 1984. "The Need for Nuclear Innovation." *Technology Review* February/March: 45–48.

Lidsky, L.M. 1984. "The Reactor of the Future?" *Technology Review* February/March: 52–56.

Lippman, T.W. 1990. "Bechtel, Westinghouse Buy into Nuclear Plant." *The Washington Post*, July 20, B1.

MacKenzie, J.J. 1984. "Finessing the Risks of Nuclear Power." *Technology Review*, February/March: 34–39.

Okrent, D., and D.W. Moeller. 1981. "Implications for Reactor Safety of the Accident at Three Mile Island, Unit 2." *Annual Review of Energy* 6: 43–88.

Thompson, T., and T.W. Lippman. 1990. "Court Orders NRC to Set Training Standards." *The Washington Post*, April 10, A25.

Wagner, H.N., and L.E. Ketchum. 1989. *Living With Radiation: The Risk, The Promise*. Baltimore: Johns Hopkins University Press.

Wald, M.L. 1990. "Barriers are Seen to Reviving the Nuclear Industry." *New York Times*, October 8.

Weinberg, A.M., et al. 1985. *The Second Nuclear Era: A New Start for Nuclear Power*. New York: Praeger.

Chapter 8

On the Public Perception of Nuclear Risk

MORRIS FARR

I have very little disagreement with the point of view expressed by Asselstine et al. (in Chapter 7 of this volume). I agree that we do have significant problems associated with nuclear power that need to be addressed. I agree that there is research which is needed to improve designs. I think there is risk associated with nuclear power, that a process for quantifying that risk exists, and that this task needs to be accomplished. The basic reason for quantifying risk is indicated by a set of words which are heard repeatedly, "public acceptance" and "public confidence." These words that have so much to do with the political and public attitude toward nuclear power, are the major reason why we need to continue quantification of the risks. My understanding is that most of the studies of risks have included assumptions which were quite conservative and as we continue the process of quantification, we are going to find that nuclear power is actually going to appear more attractive, not less.

I begin with the assumption that the real problems associated with nuclear power basically have to do with politics and with public attitude. One major specific problem is quite obvious and Asselstine et al. are correct in the importance attached to waste disposal. NIMBY (Not In My Backyard) politics has become one of the great emotional issues of our times, and some politicians have not been able to resist the opportunity to espouse extraordinarily simple solutions. Whether the facility being sited is a prison, a city dump, a nuclear waste repository, or a half-way house for teenagers, the problem is similar. When these issues, which have emotional connotations arise, it is very difficult to find a politician who will be courageous enough to say, "We should site this thing in my neighborhood." Of course, some simple solutions exist. One can simply identify the district of the minority party member who has the least seniority and determine that the facility will be located there. This process is easily understandable, but when nuclear waste disposal is at stake, I realize that the public would feel better served if a more technically oriented and rational process were employed. However, that presents a problem in itself. How do we obtain a technically oriented and rational discussion? Clearly, one of the greatest difficulties of the nuclear industry is the constant battle of explaining and defining terms and trying to find words which do not themselves raise negative

M. Waterstone (ed.), Risk and Society: The Interaction of Science, Technology and Public Policy, 121–124.

emotions. If nuclear power is to be a feasible energy alternative, there are several concepts that must be clarified or eliminated from our collective dialogue.

The first concept is radiation. Radiation is a word which has a fairly precise scientific definition. Nuclear scientists are able to talk about radiation. They can discuss alphas, betas, or gammas, and concepts like dose rates or exposure limits are not particularly frightening because they can be quantified and given meaning. However, the moment the word "radiation" is used with many members of our society, we find that the word itself generates a fear reaction, and as soon as it enters conversation the rational part of the discussion is over. The solution to this problem is not clear, but the word has to be demystified. Instead of something that has an almost magical quality, we must return to its true meaning: a scientific definition of a physical phenomenon that can be quantified and discussed rationally. Needless to say, many other words also fit this description.

Another concept which must be clarified, is embodied in the word "safe." The reason is obvious. There is no such thing as safe. Nuclear plants are not safe, airplanes are not safe, and automobiles are not safe. Even our homes are not safe. There is a finite possibility that on a normal morning, preparing for my day's activities, that I could experience a lethal shock from my coffee maker. There is nothing which we can create in our technological society which is absolutely safe. We have to get rid of that concept and get it out of the public perception, because the only thing that is reasonable to discuss is risk and whether or not that risk is acceptable. That is the idea which must be communicated to the public and which the public must come to understand. All we are doing is calculating our odds and deciding whether or not those odds are acceptable.

Clearly, there is no quick and easy way to correct these gaps in the public understanding of words. The most obvious approach is to examine the education of our children. Radiation does not seem nearly so abstract and magical to the student who has participated in a counting experiment with a radioactive isotope. Similarly, statistical concepts can be understood by young students. Many young people are talented poker players by the time they graduate from high school and have a finely tuned knowledge of the nature of risk and the impossibility of a "safe" bet. While the payoff may be some time in coming, attention to the general problem of scientific and mathematical literacy is something our educational establishment must address.

A third concept which we have to rectify or abandon, is that there is any cheap, quick fix for the problems of energy production in a technological society. When news broke of so-called "cold fusion," I had the experience of people saying, "Isn't this great! Our energy problems are going to be solved." Statements like these demonstrate a naivete about technological progress which is saddening. As a society we must rid ourselves of the idea that there is anything we can do about energy that is going to be cheap or easy. We have to consider the problems associated with not just nuclear technology, but with all the technologies and compare them. Only in that comparison will it be possible to make rational decisions about what is best for society.

Once again, education is the long term solution to the problem of scientific naivete. In the meantime, something like crash courses for journalists may be in order. Much of the public sense that "quick fixes" are available comes from media that seem to report only the spectacular and seldom do a good job of portraying the variety of opinion that usually exists within the technical community.

Perhaps the greatest problems are that many people are very comfortable with their myths and there are others who have made a career out of appealing to fear-filled myths. These myths include the "magical" dangers of radiation, the danger of nuclear power, and the prospect of an "easy way to cure our energy problems." The reason these myths persist is the technological naivete, even illiteracy of a large part of our society.

One possible solution involves an increased awareness of STS. STS is a relatively new acronym that stands for Science, Technology and Society. In the present context it refers to academic programs which are designed to increase the overall awareness and knowledge of the public about technology, the problems which accompany it, and the interaction of science and technology with society. Perhaps the most important thing we can do in order to promote a rational, decision-making process about nuclear energy or any other technological enterprise is to change the way we educate students. That has to happen at the universities, in the public schools, at every level of education.

The STS approach to education generates many questions: Do we want to start at the top and educate graduate students who have an academic interest in society, technology and science? Do we want to start earlier and educate undergraduates? Or do we want to attack the problem at the high school and junior high level with information about civic responsibilities and the need to understand technological problems and the solutions that technology can provide? My bias is that we should introduce STS programs at the earliest possible opportunity. Of course, this requires a significant change in the education of educators. Like all great problems, the only thing that is clear is that new approaches are necessary and must be tried.

Personally, I believe that the risks of nuclear power are acceptable. I also understand that many people find that a difficult position to accept in the aftermath of the incidents at Three Mile Island (TMI) and Chernobyl.

TMI was probably the most expensive non-accident in history. Upon reflection, we realize that no one was injured and the effects of the accident are not measurable. Some radiation was released but the effect on the surrounding population was less than if the exposed individuals had made an airplane flight to Denver. TMI was a good thing for a few people in the sense that many scientists and engineers are making a very good living from the analysis and clean-up that has taken place since the event. However, much more important than the financial fate of this relatively small group is the fact that the investigation represents a very positive sign about the nuclear industry. Progress occurs only when mistakes are understood and corrected.

Chernobyl was somewhat different and was, in fact, a disaster. People were killed, the area around Chernobyl has been devastated, and no sane person would claim that it was anything but a disaster. However, it is at least arguable, even after Chernobyl, that a comparison of risks to other ways of generating electrical energy could still indicate that nuclear power is not so bad after all.

For example, if our comparison is to coal-fired power, many questions arise. How many human beings die each year mining the millions of tons of coal required for coal-fired power? What are the long-term effects on public health of combustion by-products? How many problems will be generated by the emission of carbon dioxide to the atmosphere? Perhaps because these problems do not strike with the dramatic suddenness of a nuclear disaster, the debate on these issues is not nearly as emotional and productive of headlines. There are many little disasters associated with all the other ways of generating electrical energy and if we are going to be rational about evaluating power sources we have to look not just at the spectacular accident, but at all these other effects.

From airplane crashes to pipeline spills, disasters involving high technology have happened before and will happen again. The only good thing about a disaster is that it does offer an opportunity for learning. Typically, engineers and scientists assign fault to things like design flaws, inadequate materials, or human frailties. Lessons may be learned and changes may be made. This is normal and natural and the source of much human progress. The key word above is "rational." Regrettably, the thoughtful reactions of responsible technologists, coming long after the fact, do not command the same attention as the emotional response at the moment. Once again, education and thoughtful reporting are the only long-term solutions.

It would be wonderful if some easy and readily-apparent method of communicating statistical risks to the general public were available. Clearly, this is not the case and the only solution is the much more long-term process of educating the public about statistical processes, demystifying the calculations of risk specialists, and informing the public about the statistical nature of risk associated with our daily activities.

Ultimately society must and will resolve the debate about nuclear power and the only real question is about the nature of that debate. At present, the debate is often strident and laden with strongly-held emotions on both sides. This is the stuff of which the human drama is composed and the debate is a "natural" for news media reports. No reporter needs to worry about a nuclear power debate being "just another story" and the rhetoric and anecdotal information can be easily composed into a news report which will grab a headline or the top billing on the evening news. These problems will not disappear, so we have no choice but to continue the difficult process of education and retain our faith that eventually reason triumphs over emotion and fear. We must agree with Thomas Jefferson, who wrote:

> I know no safe depository of the ultimate powers of society but the people themselves; and if we think them not enlightened enough to exercise their control with a wholesome discretion, the remedy is not to take it away from them, but to inform their discretion by education.

Chapter 9

Nuclear Power: Is it Worth the Risk?

RONALD D. MILO

The Fundamental Issue

In considering the ethical implications of the use of nuclear power, the chief question to be addressed is this: Can the use of nuclear power be justified, given that it involves serious risks of harm to human beings and their environment? Are the risks involved in the use of nuclear power morally acceptable?

I believe that this is also the fundamental question that social planners and policy makers must address. Asselstine et al. point out that, because of the nuclear accidents at Three Mile Island and Chernobyl, among other things, "the public does not support a commitment to new nuclear plants at this time" (Chapter 7, this volume). They ask what can be done to restore this confidence, and then go on to suggest seven factors that need to be addressed in order to restore it. But before we address ourselves to this question, should we not first assure ourselves that this confidence ought to be restored? And to answer this question we need to ask ourselves whether the risks posed by the use of nuclear power are acceptable risks.

No doubt the general public 's lack of confidence in the safety of nuclear power is based on many misconceptions (see the previous chapter). As psychologists and sociologists have pointed out, people tend to be overimpressed with and horrified by descriptions of worst-case scenarios. Consequently, they are unable to discount the importance they attach to these worst possible outcomes by noting that they are also extremely improbable. As studies indicate, fear causes people to blur the distinction between what is remotely possible and what is probable, which makes it difficult to have open and objective discussions of the dangers posed by a nuclear accident or radiation release (Slovic and Fischhoff, 1983; Cohen, 1974).

On the other hand, evidence shows that experts are also often unreliable predictors of the likelihood of bad consequences. Studies have shown that they tend to be overconfident about their ability to make such predictions. When experts make such estimates they often set upper and lower confidence bounds, claiming that their is a 98 percent chance that the true value lies between them. But as one study points out, experiments reveal that, "rather than 2 percent of

M. Waterstone (ed.), Risk and Society: The Interaction of Science, Technology and Public Policy, 125–134.

the true values falling outside the 98 percent confidence bounds, 20–50 percent do" (Slovic and Fischhoff, 1983).

Another study indicates that such procedures typically resulted in specifying bounds that were much too narrow (Hynes and Vanmarcke, 1976). And it has been pointed out that a multimillion dollar study by the Nuclear Regulatory Commission, conducted in 1975 in order to assess the probability of a core melt down "used the very procedure for setting confidence bounds that has been found in experiments to produce the highest degree of overconfidence" (Slovic and Fischhoff, 1983: 118).

Of course, the mere fact that nuclear power poses serious risks of harm does not mean that its use cannot be justified. For nuclear power also provides us with important benefits. We commonly think that if certain benefits can be obtained only by taking certain risks and if these benefits are very important to us, these risks may well be justified. For example, heart by-pass surgery involves some risk of death as an immediate consequence of the surgery, but it also provides a benefit of vital importance to the patient – prolonged life and improved quality of life. What makes the risk acceptable is 1) the fact that the benefit is considered to have great value, and 2) the fact that, although the magnitude of the harm being risked (death) is great, the probability of its occurring is (given modern medical skills) rather small. A third factor to be taken into account is that the alternatives to the surgery – e.g., forgoing surgery and relying on medication – also involve risks (of heart attack, for example) and risks that are less acceptable than those involved in the surgery. Thus, the assessment of a given risk is no simple matter.

Risk Assessment

Indeed, the study of risk assessment has now become the object of a distinct intellectual discipline (Lowrance, 1976; Rowe, 1977; Kates, 1978; Schwing and Albers, 1980). These studies have led to the formulation of a number of general principles for assessing risks. In general – and putting it abstractly – we can say that the acceptability of a risk is to be assessed by weighing the positive value of the benefit to be gained against the negative value of the risk taken. The negative value of the risk taken is a function of two things: the magnitude of the harm being risked and the probability of its occurrence. If we can assign a numerical value to each of these, the value of the risk will be equal to the product of these two numerical values. Of course, in most cases we are unable to assign any precise numerical value either to the magnitude of the harm being risked or to the probability of its occurrence. In the case of heart by-pass surgery we may be able to assign a rough numerical value to the probability of death occurring as a consequence, if we have statistical information on enough previous instances of this kind of surgery. But in attempting to assign a numerical value to the benefits of surgery, the best we can do is perhaps to rank the prolonged life and improved quality of life resulting from the surgery on a scale from 1 to 10 in comparison with other sorts of benefits for which we might consider it worth taking comparable risks (for example, having one's sight restored, preventing the death of a loved one, or making an important scientific discovery).

In assessing the risks involved in the use of nuclear power, we must take these same factors into consideration: 1) What are the benefits to be gained from nuclear power, and how important are they, especially in comparison with other benefits for which we might be willing to take risks? 2a) What are the risks involved? And what are the magnitudes of the harms being risked in comparison with other sorts of possible harms? 2b) What is the likelihood that each of these harms will occur? (And, we might also ask, what is our basis for assigning these probability values?) 3) What are the alternatives to the use of nuclear power in meeting our energy needs, and what are the costs and risks associated with each of these alternatives? Do the alternatives also involve risks? If so, how does the acceptability of these risks compare with the risks involved in nuclear power?

The Benefits of Nuclear Power

There is no question that the availability of nuclear power increases our energy resources, and hence, given the scarcity of the latter, is highly beneficial. Indeed, some might argue that it is not only beneficial but necessary – that we cannot do without it, even if it is unsafe. But surely the latter view is too strong. Given our current energy crisis, the utility of nuclear power is certainly very great. Whether nuclear power is necessary, however, depends on what our alternatives are and whether these are acceptable.

The utility of nuclear power is well enough established by the fact that we are gradually running out of renewable fossil fuels, such as petroleum, natural gas, and coal. Currently, almost 90 percent of U.S. energy needs are satisfied by these latter, fossil fuels (Gould, 1983: 16–17). At current rates of consumption petroleum and natural gas reserves are expected to last no longer than a few decades, and coal reserves no more than a couple of centuries. Obviously there is a need for new energy resources, and the use of nuclear power is one way of meeting this need. But whether or not the use of nuclear power is essential depends on what other energy resources are available to us, as well as our ability to make do with less energy. Of course, even if it is possible to do without nuclear energy, it may not be desirable. There may be unacceptable risks (such as acid rain and global warming) associated with the use of alternative sources of energy. And we may not wish to pay the price of a lower standard of living entailed by simply cutting back on energy consumption.

The Risks Posed by Nuclear Power

The two most serious risks involved in the use of nuclear power appear to be: 1) the danger of a nuclear accident during the operation of nuclear plants; and 2) the possibility that efforts to contain stored nuclear waste materials will prove inadequate, resulting in the leakage of radioactive materials. There are, of course, other risks as well. For example, accidental radiation exposure to workers may occur during the normal operation of nuclear plants, and also in connection with the production and transportation of nuclear materials. But the two major risks I have mentioned have been the cause of the most serious concerns, and I want to focus particularly on the second of these: waste disposal.

Radioactive waste materials that are a byproduct of the production of nuclear energy include "the tailing from uranium mining, effluents from fuel enrichment and fabrication plants, the fission and transuranic byproducts of fission reactions, the ordinary metals made radioactive in nuclear reactors by nuclear bombardment, and all the materials in all segments of the nuclear fuel cycle that become contaminated through contact with radioactive elements" (Gould, 1983: 1). These waste materials have varying levels of radioactivity. Some are rather short-lived and decay relatively quickly. But others will pose a danger not only for us, but for our great-great-grandchildren as well (and even more distant future generations). Presently only low-level wastes are being disposed of on a semipermanent basis. The more highly radioactive wastes are now being stored in temporary facilities, pending the development of permanent disposal sites. Current plans call for embedding these materials in glass, then placing them in steel containers, and depositing these in deep salt beds (Kuhlman, 1976).

Unfortunately, the success of strategies already employed for dealing with low-level wastes does not bode well for current plans to dispose of high-level wastes. It was thought by engineers that the low-level wastes could safely be disposed of in shallow land burial sites, but it was discovered that some of these wastes were able to migrate away from these sites. It was also found that some of the steel tanks used to temporarily store liquid high-level wastes have leaked (Gould, 1983: 3). This has led to some skepticism about whether the current plans for permanent storage of the high-level wastes are really going to make them safe. The steel containers are expected to corrode after a few hundred years, and some fear that the geologic barriers counted on to prevent groundwater from reaching and leaching the radioactive wastes may fail – perhaps as a result of earthquakes or other unanticipated events (De Marsily et al., 1977). In the event of such an occurrence, the consequences could be as harmful as those from a nuclear explosion.

The mere fact that the disposal of nuclear waste materials poses a risk of such a magnitude does not mean, however, that this risk must be unacceptable. As we have seen, that will depend also on how probable (or improbable) this event is. It will also depend on what alternatives to the use of nuclear energy there are, and what costs or risks are associated with these.

What Are the Alternatives?

The major alternatives to the use of nuclear power appear to be these: 1) We can continue to use energy at the current rate and substitute other energy sources to meet the needs currently expected to be met by nuclear power; 2) We can curtail our consumption of energy in order to obviate the need for nuclear energy; or 3) We can combine conservation efforts with some package of more desirable alternative energy sources.

If we wish to continue to use energy at the current rate, then, it seems, our only alternative is to rely more heavily on the use of nonrenewable fossil fuels. We do not presently have the technology to enable us to derive all the energy we need from other energy sources – such as solar, geothermal, and hydroelectric. Unfortunately, the use of fossil fuels carries with it its own hazards. Moreover, these hazards are not just remote possibilities. They are highly probable, and some are virtually certain to ensue. Air pollution and acid rain are already apparent results. And the increasing concentrations of carbon dioxide resulting from the use of such fuels appears to be having a "greenhouse effect" that is causing a gradual warming trend leading to other adverse climatic changes (see Chapter 2, this volume). The latter is an especially serious consequence, whose harmfulness certainly rivals and may even exceed the harms risked through the use of nuclear power. These claims about fossil fuels are especially true in the case of the most abundantly available fossil fuel – coal, whose use also carries with it the (not merely remotely possible) dangers associated with mining. Thus, the first alternative posed above does not seem to be a very viable one.

The consequences of obviating the need for the use of nuclear power by curtailing our use of energy are not harmful or dangerous, but many people would certainly find them unpleasant. Conservation measures will at least involve certain inconveniences and may even entail some rather significant changes in lifestyle. Some have suggested that it would involve restructuring our present forms of industrial societies in order to make them less dependent on transportation and high energy consuming industries (Woodhouse, 1972; Schumacher, 1973; Lovins, 1977). And some have suggested that it might entail reverting to an essentially agrarian society, since "even a restructured industrial society would require more energy than can be provided by agriculture alone" (Gould, 1983: 18). When one considers all of this, the second alternative mentioned above also begins to lose its attractiveness.

The most promising alternative seems to be some combination of conservation measures together with a greater use of non-fossil energy sources, such as hydroelectric, geothermal and solar energy. But even with rather significant, and therefore rather costly, conservation measures, it seems doubtful that these other energy sources could carry much of the total energy load that would still be required – if we are not to drastically change our style (and perhaps standard) of living. Some have concluded, however, that "there would be no need for nuclear energy in the foreseeable future were we to pursue modest efforts at energy conservation that are technologically feasible, economically profitable and broadly consistent with present lifestyles" (Goodin, 1978: 31). Whether one agrees with this or not will depend on what one considers to be "broadly consistent with present lifestyles" and what sort of lifestyle one considers acceptable.

Thus, the comparative assessment of the risks associated with the use of nuclear power poses some real practical and moral dilemmas for us. It is not easy to see how to resolve them.

The Complexity and Subjectivity of Risk Assessments

It is important to realize also just how complicated the assessment of risks is. Moreover, these assessments are much more subjective and much less determined by any hard empirical data than experts are usually willing to admit. We may begin by noting that the assessment of a given risk will depend on two kinds of judgments. First, there are various kinds of scientific judgments. These include judgments about the possible consequences of certain procedures and untoward events (e.g., nuclear accidents and radiation leakage), and judgments about the likelihood that each of these possible consequences will occur. Although it may appear that these judgments are quite objective, because they are based on empirical data and other scientifically established facts, it turns out that they are often based on highly contentious and subjective interpretations of the data. Moreover, estimates of the likelihood of bad consequences are typically seriously underdetermined by statistical information and hence really deserve nothing more than the label, "guesstimate."

Another reason why risk assessments cannot be determined by empirical data alone is that they also involve value judgments. These include judgments both about the relative seriousness or badness of the harms being risked and judgments about the comparative importance of the benefits to be purchased with these risks. For example, whether a particular person (or group of persons) thinks that the risks involved in the use of nuclear power are acceptable will depend not only on their judgments of the likelihood and relative badness of a nuclear accident or radiation exposure due to inadequate containment of nuclear waste materials, but also on how important they take the benefits made possible by nuclear power to be. They may agree in their estimation of the relative badness of these events, but disagree about whether these risks are worth taking simply because they make different evaluations of the benefits. For example, they may agree that, since the only currently feasible alternative to the use of nuclear power for meeting our energy needs is the use of fossil fuels, and that since these are known to have very harmful effects on the environment (such as acid rain and global warming), forgoing the use of nuclear power will mean that we must either accept other bad consequences or curtail our appetite for energy. Weighing the comparative badness of the almost certain harmful effects of fossil fuels against the possible but unlikely bad effects risked in using nuclear power is an extremely difficult and somewhat subjective matter.

If we elect to avoid this dilemma by choosing to curtail energy consumption, this will mean that we will have to change our lifestyles and perhaps even to reduce our standard of living somewhat. But while agreeing that this is the only acceptable alternative, some may think that the risks associated with nuclear power are worth taking because they find the required changes in lifestyle that would otherwise be required to be unacceptable, whereas others will conclude that the risks are not worth taking because they find giving up such things as being able to commute to work in one's personal car instead of using public transportation to be not so onerous.

As many moral philosophers have regretfully come to conclude, disagreements of this sort often reflect basic differences in fundamental attitudes and preferences – differences that persist despite the fact that everyone knows and agrees on the facts. So this is another important respect in which risk assessments are underdetermined by empirical data.

These comments about the inherent complexity and subjectivity of risk assessments are well illustrated when we consider what would be involved in making an accurate assessment of the risks associated with nuclear waste disposal. The danger here is that efforts to contain the radioactive waste materials that are a natural byproduct of producing nuclear energy will prove inadequate. As we noted earlier, the most dangerous of these materials are being maintained in temporary surface storage facilities, pending decisions on their final disposition. The most feasible plan for the latter involves embedding these wastes in glass, encasing the glass in steel containers, and then storing them in stable geologic deposits.

Estimating the risks involved in this process entails, first of all, anticipating the possible ways in which these containment efforts might be frustrated by such things as unanticipated groundwater leakage, earthquakes, sabotage, meteorite impacts, etc., and then predicting the likelihood of each of these kinds of events. As we noted, estimates of the probability of these kinds of events are only very weakly determined by empirical data. For some kinds of events, such as meteorite impacts, for example, there might be some relevant historical and statistical information. For others, such as sabotage, there may be almost nothing to go on.

We must also try to anticipate the various kinds of harms and damage that will ensue in the case of each of these possible but unlikely causes of radiation release. How many deaths will occur? What about illnesses and genetic defects? We will also need to consider property damage and long-term damage to the physical environment rendering it unfit for use. Here we are required to make judgments of the comparative badness of each of these kinds of losses – judgments which we have noted to be inherently subjective. Moreover, if we wish to make any tolerably precise assessment of the risk, we will have to quantify each of these bad consequences by, for example, attaching a dollar amount to each kind of loss – an admittedly arbitrary kind of thing to do. We would need to do this because, as we have seen, the negative value of the risk is determined by multiplying, for each loss, the probability of that bad event's occurring times the magnitude of the badness of the loss.

But the complexities and indeterminacies do not stop here. Having determined the negative value of the risk, we must now weigh this against the positive value of the benefit for which the risk is taken. And we must consider not only the alternative of simply forgoing the benefit (in this case having a greater capacity to meet our energy needs), we must also consider alternative means of realizing this benefit (such as greater use of fossil fuels). However, when one realizes that the use of fossil fuels has bad consequences of its own (as already elaborated), one is faced with a difficult value judgment: Should we cut back on energy consumption and accept a somewhat less desirable lifestyle as a consequence, or should we take the gamble and continue to live in the style to which we have grown accustomed? Individuals may find it difficult to answer this question for themselves. Gaining any kind of social consensus on an answer is likely to prove even more difficult.

Decision Making and Public Policy

Having now observed the inherent complexity and subjectivity of risk assessments, what implications does this have for social policy makers? What is the wise and prudent policy for us to adopt collectively? What are the ethical implications of making policy decisions that have possible harmful consequences for others (not a party to these decisions)?

What is the prudent policy for us collectively? One lesson that might be drawn from this discussion is that we should be wary of placing too much confidence in risk assessments made by those considered to be experts. But then how should these estimates be treated? Presumably, with caution. It may be best to try to err on the side of caution in determining whether such risks are acceptable. One way of doing this would be to revise our estimate of the degree of comparative benefit required in order to make a given risk acceptable. Since the worst-case scenario is so disastrous, perhaps we should make sure that we really cannot do without the energy from nuclear power, even if it does involve some inconveniences and changes in lifestyle. We must recognize also that even though the possibility of such a disaster is extremely remote its magnitude is also very great. Thus, as one opponent of the use of nuclear power puts it: "Even the most dedicated defender of nuclear power must concede that leakage poisoning many people is a possibility, however remoted a one he may think it to be. Since the catastrophe in question is of sizable proportions, it is worth a major investment for us to make sure it does not occur" (Goodin, 1978: 33).

One important ethical question that is raised by these considerations is this: Given the inherent complexity and subjectivity involved in such risk assessments and given, consequently, the low degree of confidence that one is warranted in placing in them, do social planners have the right to impose policies based on their own risk assessments on an unwilling or reluctant public? Indeed, should they even try to sell the public on policies based on such questionable and contentious foundations?

It is generally agreed by moral thinkers that risks of harm ought not to be imposed on others without their consent – even if one judges that it is in their own best interests. This means that social policies that involve incurring serious risks of harm to people must be determined in a democratic fashion, and only with the full consent of those affected by it. This means also that those in charge of formulating and suggesting policies must make every effort to ensure that those affected are fully and honestly informed about the nature of the risk involved. Honesty requires the avoidance of euphemisms, such as "operational incidents" to refer to nuclear accidents that pose dangers. And it also means, of course, that all such accidents are publicly reported in an unbiased manner.

A further ethical question concerns the rights of future generations. It may well be that the disposal techniques and precautions proposed by engineers for dealing with high-level and very long lasting radioactive materials will protect their contemporaries. But what about the future generations? The radioactivity of some of the materials can last for thousands of years, while the canisters designed to contain them are generally expected to corrode after only a few hundred years. Thus, social policy makers need to be mindful that in choosing to take a risk they are taking a gamble not so much for themselves as for their great-grandchildren and more distant descendants. Moreover, they are choosing to impose a risk on others for the sake of their own benefit; and choosing to impose this risk on future generations without their consent. This is a clear issue of intergenerational justice. It is difficult to see how we could have the right to impose such a risk on future generations. On this score I am inclined to agree with Robert Goodin when he concludes that "the gain from cheap and clean energy now is nowise commensurable to the risk of radiation poisoning we are creating for future generations" (Goodin, 1978: 33).

Thus, even if one could successfully argue that accepting the risks associated with nuclear power represents a wise and prudent policy for those who choose it for themselves, it does not seem right for them to impose these risks on others. And, of course, the argument for the wisdom of choosing to take such a risk for oneself is itself highly questionable. Given the catastrophic nature of the harms being risked, it is not at all obvious, and in fact quite dubious, that the benefits to be gained outweigh the value of the risks.

References

Cohen, B.L. 1974. "Perspectives on the Nuclear Debate." *Bulletin of the Atomic Scientists* 30(9): 35–39.

De Marsily, C., et al. 1977. "Nuclear Waste Disposal: Can the Geologist Guarantee Isolation?" *Science* 196 (August 5): 519–827.

Goodin, R.E. 1978. "Uncertainty as an Excuse for Cheating Our Children: The Case of Nuclear Wastes." *Policy Sciences* 10: 25–43.

Gould, L.C. 1983. "The Radioactive Waste Management Problem." In C.A. Walker, L.C. Gould and E.J. Woodhouse, eds., *Too Hot to Handle*. New Haven: Yale University Press.

Hynes, M., and E. Vanmarcke. 1976. "Reliability of Embankment Performance Predictions." *Proceedings of the ASCE Engineering Mechanics Division Specialty Conference*. Waterloo, Ontario, Canada: University of Waterloo Press.

Kates, R.W. 1978. *Risk Assessment of Environmental Hazard*. New York: Wiley.

Kuhlman, C.W. 1976. "ERDA Waste Management Program." In *Proceedings of Conference on Public and Policy Issues in Nuclear Waste Management*. Chicago, October 27–29. Washington, D.C.: Harrison Associates.

Lovins, A.B. 1977. *Soft Energy Paths: Toward a Durable Peace*. Cambridge, MA: Ballinger.

Lowrance, W.W. 1976. *Of Acceptable Risk*. Los Altos, CA: W. Kaufmann.

Rowe, W.D. 1977. *An Anatomy of Risk*. New York: Harper & Row.

Schumacher, E.F. 1973. *Small is Beautiful: Economics as if People Mattered*. New York: Harper & Row.

Schwing, R.C., and W.A. Albers. eds. 1980. *Societal Risk Assessment: How Safe is Safe Enough?* New York: Plenum.

Slovic, P., and Fischhoff, B. 1983. "How Safe is Safe Enough? Determinants of Perceived and Acceptable Risk." In C.A. Walker, L.C. Gould and E.J. Woodhouse, eds., *Too Hot to Handle*. New Haven: Yale University Press, pp. 112–150.

Woodhouse, E.J. 1972. "Revising the Future of the Third World: An Ecological Perspective on Development." *World Politics*, October 1–33.

PART IV

Setting Standards for Air Quality

Science and public policy intersect directly when policy makers confront the difficult problem of setting standards to protect public health, safety and welfare. Often it is possible for scientists to identify potential hazards, and, to some extent, to quantify the risk of occurrence and the possible severity of the consequences. However, scientists (acting in their capacity as scientists) cannot answer the fundamental question regarding the degree to which society should be protected against such potential hazards. The answer to this question must be based upon an assessment of the relevant social values; the notion of "acceptable" risk.

When public health standards and regulations are being developed, all of the tradeoffs between the potential benefits of a new substance, process or technology must be weighed against the possible risks and adverse consequences. The costs (economic and otherwise) of achieving a desired degree of public protection must be compared with the benefits obtained (improved public health or safety) or foregone (i.e., the benefits that might have accompanied a particular product or technology).

The chapters in this section explore this issue through an examination of the complexities involved in setting standards for ambient air quality. This standard setting process encompasses all of the difficulties encountered by policy makers and analysts alike when trying to answer the deceptively simple question "How safe is safe?"

Chapter 10 presents Milton Russell's examination of the choices involved in controlling the pervasive problem of ozone pollution. Dr. Russell begins by examining the origins and effects of tropospheric ozone. He then places these comments in the context of our regulatory approach for controlling ozone (and other air pollution) through the Federal Clean Air Act. Finally, Dr. Russell identifies the targeted health standards for ozone, where we stand in attaining those goals, and what would be required for achievement.

In Chapter 11, Michael Lebowitz offers an assessment of the role of respiratory and other related sciences in providing the information base necessary for developing the kinds of standards described by Dr. Russell. Professor Lebowitz indicates the drawbacks of current approaches and suggests several strategies that would improve this process.

In the final chapter David Baron, an environmental, public interest attorney, takes a hard and critical look at the uses and abuses of risk assessment, especially as applied to the setting of public health standards. Mr. Baron indicates the ways in which risk assessment can be manipulated to achieve particular, desired ends. He cautions against an overly naive faith in risk analysis or the purportedly "objective" science upon which such analyses are based.

Chapter 10

Ozone Pollution: The Hard Choices[1]

MILTON RUSSELL

Introduction

Ozone, the major component of photochemical smog, is arguably the most intractable political/economic/environmental problem facing the United States today. After almost two decades of effort, in 1987 sixty-eight areas of the country still failed to meet the ozone ambient air quality standard set by the Clean Air Act (CAA).[2] The hot, dry summer of 1988 left still more areas out of compliance, and recorded higher levels of violation in many of those areas not in compliance before. Complicating the matter further, new research findings suggest that the health effects of ozone have additional dimensions not contemplated in the setting of the original standard for ambient air quality. Moreover, other research calls into question the strategy for controlling ozone on which previous actions have been based.

The legislative situation has compounded the intense interest in the ozone issue. A deadline for meeting the standard, which had already been extended twice before, was December 31, 1987. Under the law, EPA had to impose substantial sanctions against some areas after that date. Congress, however, extended that deadline yet again until August 31, 1988, to enable it to decide what amendments to the CAA were appropriate given the existing conditions.[3] At the time of this writing several approaches are under consideration, but all of them build on the basic structure of the existing CAA.[4] Consideration of the ozone problem is further complicated because the reauthorization of the CAA also contemplates such other contentious issues as acid rain and toxic air pollutants.

The difficulty in deciding what to do is understandable. As will be discussed further below, controlling ozone on current understanding and in the CAA framework will be extraordinarily difficult, expensive and disruptive. The technical feasibility of achieving the goals of the CAA is itself in question. Ozone control has significantly different impacts on different sectors of the country and on different economic interests. It raises fundamental issues of federalism between the Federal government and states, and among political jurisdictions. And new research results not yet fully assimilated are calling into

M. Waterstone (ed.), Risk and Society: The Interaction of Science, Technology and Public Policy, 139–164.

question the received wisdom of what ozone regime is required to protect public health and what control strategies might be most effective in achieving it.

At the same time, the pressures on Congress are all to act now and not to do so can be painted as irresponsible. The public is frustrated that the promise of the CAA that all citizens will be protected from adverse health effects has not been met. Yet, the thesis of this chapter is that no legislative action should be taken at this time; that it would indeed be irresponsible to set the nation on the course envisioned by any of the proposals now being considered. Instead, current policy, which is at the least preventing the underlying problem from getting worse, should be continued until needed research is done, new data are assimilated, and a fuller public debate on the goals and framework of air quality protection takes place.

So far the ozone debate has been most notable for what is missing. For example, there has been little public discussion of the actual number of people who are at risk, to what health effects, of what level of concern, for what number of hours. There has been less discussion of alternative uses of some of the direct national resource expenditure (above current levels) of $100 to $150 billion by the year 2000, and over $20 billion per year thereafter, found in the proposals being considered. Much less has there been consideration of balancing the personal, regional, and national sacrifices that would be entailed against what would be gained. A matter of this importance deserves a debate that includes such issues.

In support of the thesis that legislation should be delayed, this chapter provides a non-technical summary review of what is known about the source of ozone and of the health and environmental consequences of exposure to it. This is followed by a discussion of the framework of the CAA and what is meant by meeting its provisions. The recent status of ozone air quality is then examined, followed by a discussion of what actions appear to be required to meet the provisions of the CAA. Based on this discussion, a different conception of the appropriate goal of air quality legislation is suggested. This is one based on the premise that it is essential to consider the broader implications of attaining different levels of air quality in different locations, rather than on the premise that government should act to prevent all adverse effects whatever the ancillary consequences. A sketch of such an approach is given, and the research necessary to support it suggested. In an Epilogue, an outline of an alternative structure for ozone control is presented.

Origin of Ozone

Atmospheric ozone is formed when reactive hydrocarbons [from volatile organic compounds (VOCs)] are mixed with nitrogen oxides (NOX) in the presence of sunlight. Recent information has tentatively implicated carbon monoxide (CO) as a precursor of ozone as well. The atmospheric chemistry is complex, but the prevailing opinion on which regulation has been based has been that in most locations the reaction is hydrocarbon limited. This view is subject to increasing skepticism.

The complexity of the atmospheric processes is illustrated by the fact that it appears that under some circumstances reducing NOX levels can exacerbate the problem. It also has been suggested that the relative contribution of control of the two gases to reducing ozone may depend on the time and location at which they are emitted within the airshed. Consequently, rather than a brute-force effort to reduce reactive hydrocarbons, a much richer, more sophisticated and site-specific control approach may be called for. Research is needed to understand the actual situation and on which to base such an approach, if it proves supported. Nonetheless, the current control strategy being pursued is directed toward reducing VOC emissions.

The sources of hydrocarbons are all but ubiquitous. Some are from natural processes, for example from plant life ranging from algae to trees. Industrial processes, refineries, and transfers of hydrocarbons (filling petroleum tanks) are one set of anthropogenic sources. Solvents, paints, dry cleaning fluids and inks are another. Emissions from incomplete combustion are yet another, as are fugitive emissions from leaks and evaporative emissions from fuel tanks. Still another source is consumer products ranging from aerosol propellants to household cleaners. The private automobile is a major contributor; it is dominant in some locations. Automobile emissions arise from tailpipe emissions, refueling, and routine untrapped evaporation from the fuel system. Refining, transporting and marketing operations to support the automobile fleet obviously make it a major source of emissions higher in the fuel chain as well.

The role played by sunlight means that ozone formation is limited to the daylight hours. Heat speeds the reaction (plus adds evaporative emissions) and therefore concentrations are high only in the summer months. Obviously, ozone is especially difficult to control in the southern United States, where intense sun and heat lead to a faster reaction of the available hydrocarbon molecules and therefore to higher concentrations. Ozone levels also rise in the absence of dispersal provided by winds. Topography is important to the extent that basin locations, such as in Los Angeles, experience air stagnation which allow concentrations of the pollutant to build. Ozone is a highly unstable molecule. With cooling temperatures and reduction of sunlight, ozone formation declines while its destruction continues. Therefore, levels tend to drop after sunset, are at their minimum in early morning, and peak at mid-afternoon. Typically, then, ozone exhibits strong locational, seasonal, weather and diurnal patterns.

The instability of the ozone molecule distinguishes this pollutant from most others of concern: it is usually present in harmful quantities only out of doors. Ozone tends to break down on contact with a surface, and therefore air moving through ventilation systems or indeed passing through window screens will largely be purged of ozone. Those persons in enclosed spaces such as buildings or automobiles are therefore protected from exposure to levels even approximating those found in open spaces.[5]

The pattern of daily sharp peaks found in isolated metropolitan areas may not be present in regional urban agglomerations such as Southern California and the Northeast Corridor. Ozone and its precursors imported from up-wind can be superimposed upon that locally produced, and since it is out of phase, may yield plateaus of elevated ozone even if it does not cause exceedances of the standard. Further, regional transport may contribute to a build-up of peak levels, and these peaks may occur outside the neighborhood where the emissions were concentrated.

One matter of non-trivial regulatory importance is that ozone formation does occur from non-anthropogenic precursors, and the same conditions that exacerbate the one affect the other. This means that allowable anthropogenic loadings must be reduced commensurately to attain any target concentration maximum. While there is no unambiguous way of determining the natural contribution at each location, a reasonable estimate of background concentrations on a one-hour daily maximum is on the order of 0.03 parts per million (ppm) to 0.05 ppm.[6] The level depends, of course, on site-specific factors, and could be higher in areas heavily forested by trees with high VOC emissions.

Effects from Ozone

Adverse health effects of ozone can be characterized briefly.[7] Ozone is a pulmonary irritant that at sufficient dose causes transitory symptoms and decrements in lung function. There may also be synergistic effects with simultaneous exposure to other pollutants. Chest discomfort and cough have been found among a fraction of normal, healthy people exercising heavily while subjected to ozone at the level of the current standard. There is epidemiological evidence for an increase in the number of asthma attacks at high ozone levels, but this has not been demonstrated under experimental conditions. Increases in hospital admissions have been reported following high ozone levels. Some evidence also exists that the observed transitory effects can bring about permanent decrements in pulmonary function. There is evidence from animal experiments at relevant doses of increased susceptibility to bacterial and viral infections, and of lung structure damage. Therefore, the observed human effects from short-term exposure, while transitory, are potentially of chronic significance, and may be cumulative.

Some sub-populations may be more susceptible to adverse effects than are others, particularly groups with pre-existing respiratory disease. Likely also more at risk are those whose pulmonary function is otherwise depressed, or who are physically stressed such as the ill or the elderly. Exercise lowers the dose levels at which any of these effects are observed. EPA staff scientists summarized the scientific findings as suggesting that for standard setting purposes under the CAA definition (discussed below) a one-hour average range of 0.08 ppm and 0.12 ppm would be appropriate, with 0.12 ppm representing the lowest observed adverse effects level for healthy, exercising subjects, but offering little margin of safety.[8]

The CAA focus on short-term exposure which resulted in the one-hour standard was based on the belief that the effects noted above were the principle ones of concern, on the expectation that high ozone levels would occur only for short periods each day, and that effects from serial exposures were independent. The realization that longer term plateaus of elevated ozone might exist in regional settings has led to examination of effects on those exposed for longer duration. This research now suggests that continued exposure for periods of six or eight hours at levels at or below those of short-term concern may have previously unsuspected adverse effects, perhaps with chronic significance. There is also suggestive evidence of a build-up of consequences from repeated short-term exposures. Consequently, further attention has been directed toward non-short-term exposure regimes, and new evidence is being reported. That evidence may be important in the design of a standard protective against adverse health effects.

As noted, the purpose of this cursory characterization of health effects from ozone is to form a predicate for the policy issues discussed below. To summarize, for policy purposes there is no reasonable doubt that even short-term exposure at some level of ozone leads to increased probability that members of the general population will experience adverse health effects. For some members of the population, in some circumstances, the threshold for such effects may be quite low – perhaps near background levels. A positive dose response is observed; the higher the pollutant level, the more pronounced the symptoms and the larger the proportion of the exposed population that will experience them. There is growing evidence that extended exposure exacerbates these effects, and perhaps may bring others, and therefore duration of exposure deserves further attention.

What is not clear, however, is what to make of these findings. The acute effects noted are often minor or sub-clinical, not life-threatening, mostly transitory, and for many people self-limiting (exercise levels can be reduced or those affected can spend more time indoors, for example). These effects are matters of concern, but arguably of a distinctly different class than those from pollutants which can cause birth defects or cancer, for example. (Note, however, that some of the VOCs that would be reduced in an effort to control ozone could have such effects.) Chronic effects, if and when demonstrated, may be of greater significance. The issue of possible public policy approaches to the existence of these health effects from ozone is addressed below.

While the health effects from ozone gather the most public attention, the harmful effects of this pollutant are of broader consequence. It is well established that at levels below those where acute health effects are observed ozone also damages agricultural plants and lowers crop yields, retards tree growth, and harms ornamental plants and shrubs. There is also evidence that ozone-related stress increases susceptibility of vegetation to drought and insect damage. There may be harmful effects on other organisms as well. Ozone damages materials such as rubber, some plastics and dyes, and paints. As the principle constituent of photochemical smog, it lowers visibility, and that can be medically significant because of its psychologically depressing effect. Decreased visibility directly degrades amenities in regions affected.[9] These non-health effects lower the quality of life of those in affected areas both directly in terms of reduced satisfactions from surroundings and indirectly through the economic losses imposed.

Structure of the CAA

The CAA sets up a complicated regulatory system predicated on the theory that the Federal government will decide what the quality of the air should be (the standard) and the states will act to attain it, based on local conditions. EPA does have authority to regulate emissions from mobile sources and the fuel system that supports transportation, and can establish national source-specific regulatory guidance for states to follow under different sections of the existing Act. At present, however, these Federal provisions are predicated mostly on matters of practicality; they affect sources not amenable to disparate local control such as mobile sources or they embody activities such as determining engineering or other parameters on a national basis that otherwise would require each jurisdiction to engage in duplicative efforts. As such, Federal authorities are designed to form only a portion of the regulatory structure to assure ambient air quality.

The responsibility to achieve requisite air quality rests squarely on the individual state or air quality region. States or subdivisions under state supervision have the duty to formulate State Implementation Plans (SIPs) for meeting the ambient air quality standards. SIPs are to encompass control actions required and see to their enforcement. They are created from monitored and modeled levels of air quality, local emissions inventories (showing how much of what pollutants are emitted where), and the expected ambient levels of pollutants that will result with different regulations in place. Presumably, if the SIP is adequate and fully complied with, there will be no violations of the standard as defined. In the terms of the CAA, the region will be "in attainment" (of the air quality standard). Because attainment is local, the SIP must be local – that is, consistent with conditions found in the metropolitan area. The CAA envisioned the need and desirability of communities to have the flexibility to accommodate both local conditions and local views of a fair sharing of the burden in accomplishing the national objective of healthy air.

The CAA provides for Federal oversight to assure that the goals of the Act are met by the states, with sanctions to be imposed by EPA in the event of non-performance. The complex system of sanctions was designed to move the area toward attainment by limiting growth in the non-complying region directly, and by imposing economic penalties until a suitable SIP is in place. Sanctions available to EPA include imposing bans on construction of large polluting sources and withdrawal of Federal support for highway and sewage treatment construction. Federal grants that fund state air quality programs can also be withheld. The ultimate "sanction" is Federal preemption – wherein after failure of the state to perform, EPA takes responsibility for determining what control actions should be taken.

EPA is given fairly explicit instructions in setting the maximum ozone concentration, the standard, from which much else follows. The CAA calls for the primary standard to protect the public health with an "adequate" margin of safety. As interpreted from the legislative history and court decisions, that goal is reached only when even "sensitive populations" (but not every person in them) are protected from "adverse" health effects with an "adequate" margin of safety; neither economic costs nor other adverse impacts of implementation can be considered in setting this primary standard. The maximum concentration in the existing standard is 0.12 ppm on a one-hour average.

The standard-setting process begins with a search of the scientific literature for evidence of health effects from a pollutant. The key question is at what level of pollution adverse effects are observed, with due attention to sensitive populations. Armed with the results of the survey of scientific knowledge, the Administrator of EPA determines what margin of safety below the demonstrated no adverse health effects level is adequate to protect public health in his or her judgment, and sets the standard.

The structure of the CAA is a binary one such that the standard is to be met for the worst air quality location monitored within an area. If the standard is not met at that location, the area as a whole is in "non-attainment." Under the current implementation of the statute, an area is in non-attainment when, for the monitor with the fourth highest separate day reading during a running three-year period, that reading is above 0.12 ppm of ozone. (Multiple excess readings on a single day are counted as one exceedance of the standard.)

There are several important implications of this structure for determining the acceptability of air quality in a region. First, a binary system of in- or out-of-attainment can lump together areas with vastly different air conditions. It discriminates neither between near-pristine areas and those which just meet the standard, nor between those which barely fail the test and those where conditions are substantially worse.

Second, aberrant weather conditions can cause an excursion above normal maximum peak levels that does not reflect the true underlying conditions in the airshed. Third, the entire airshed is characterized by the location of the monitor which registers the fourth highest reading over a three-year period. This monitor may not be located where maximum concentrations occur some year, which means that the area is falsely thought to meet air quality goals. Alternatively, the highest reading monitor may represent the actual conditions in only one small segment of the airshed, and therefore suggest unsatisfactory air quality for the whole region which in fact is experienced by only a small fraction of the population.[10]

Finally, the "out-of-attainment" designation may give a false impression of the actual number of hours of exposure experienced by even those located at the monitor with the highest reading. For example, hourly readings during the hot months of the ozone season from the "highest monitor" in the Houston area were analyzed for the period 1981–1985. Though Houston has one of the most severe ozone problems in the nation, this monitor recorded levels above 0.12 ppm only about one-half of one percent of the hours.

In general, the attainment/non-attainment dichotomy (and the data on which it is based) found in the CAA is a very poor proxy with which to summarize the actual air quality experienced by a population residing in an area. Yet, it is the measure that drives both public understanding of the situation and regulatory responses to address air quality.

Ozone Attainment Status

It is extremely difficult to characterize the trend in the ozone situation. No satisfactory summary statistic is available, and even if it were, year-to-year variations can swamp any trend over a short period. For example, the three-year period 1984–1986 compared to 1983–1985 showed 14 fewer areas in non-attainment, perhaps in part because data from the hot summer of 1983 fell out of the rolling three year calculations.[11] Data from 1987 caused a net six areas to be added to non-attainment status, and undoubtedly data from the summer of 1988 will add more areas still.[12]

Looking over a somewhat longer period, EPA reports that between 1979 and 1985 estimated hydrocarbon emissions decreased 12 percent, and the composite average of the second highest daily maximum one-hour ambient ozone values decreased by 10 percent.[13] New regulations limiting hydrocarbon emissions were imposed during this period and since, expected emissions per vehicle in the automobile fleet declined due to replacement of less controlled automobiles with more controlled ones, and other actions also were implemented. Based on these actions already taken and the VOC control strategy in place, there should be an underlying improving trend for a few years. Then the expected growth in population, economic activity, and vehicle miles driven will begin to offset the "once-and-for-all" improvements per car and per unit of industrial output being phased in and ozone levels should rise.

The actual pattern of ozone levels has exceeded those expected from the theory on which controls are based and estimates of reductions in hydrocarbon emissions, which has caused many observers to look for an explanation. Is the monitoring system too crude to measure real changes? Is the problem a faulty understanding of the atmospheric processes such that hydrocarbon control is not the means to success? Do major gaps exist in the inventory of the hydrocarbon emissions such that the base is larger than thought, and that emissions from previously uncounted sources have actually increased?[14] Or have the regulations counted on to lower VOC emissions not been implemented – or if implemented, proven less effective than anticipated in reducing reactive hydrocarbon loadings? These are among the hypotheses that need to be explored.

While the overall ozone trend and its future direction remain unclear, the existing non-attainment situation as defined by the CAA criteria does present a picture of about one-third or more of the nation's population residing in areas which do not comply with the standard.

The data in Table 1 demonstrate the diversity of ozone non-attainment conditions. (Some non-attainment areas and levels change year by year but the pattern of diversity remains.)

A few cities, with Los Angeles as the premier example, are in a class by themselves. For Los Angeles, despite controls already much more stringent than those in most of the rest of the country, the fourth highest expected reading was almost three times the standard, and exceedances would occur about two-fifths of the days. Other non-attainment areas, on the other hand, are expected to be only slightly above the standard, perhaps for no more than one or two days per year.

Another way of looking at the problem is to distinguish isolated metropolitan areas with ozone problems, such as St. Louis, Atlanta, or Cincinnati, from urban agglomerations, such as Southern California or the Northeast Corridor from Northern Virginia and Washington, D.C. up beyond Boston. Clearly the difficulty of determining emissions inventories, understanding atmospheric processes and devising technically effective and politically tractable strategies for dealing with the former is substantially less than for the megalopolis.

Attaining the Standard: What Would Be Required

As noted above, non-attainment situations are vastly different. They would require measures commensurately different in social and economic impact to rectify. For some areas, continued or perhaps intensified enforcement of existing and forthcoming VOC regulations will bring attainment, and all that is required is continued vigilance.

For other areas the story is quite different. Based on somewhat earlier data (1982–1984), an EPA study (the "FIP Study") found that the nine cities in which the problem is most severe would require hydrocarbon emission reductions of 60 to over 70 percent, thirteen more cities would require 50–60 percent reductions, and a total of 37 cities would have to reduce emissions by more than one-third.[15] Again, this study was based on a hydrocarbon strategy, though as noted above a richer mix of instruments may be called for as further understanding of atmospheric processes in each specific location is gained.

Some indication of the magnitude of the effort required can be gathered from a preliminary EPA attempt to "scope out" packages of possible hydrocarbon control measures to achieve attainment in representative cities. The "low" reduction package achieving 25 percent reduction is about what

Table 1. Ozone

Areas With 1985–87 Ozone Expected Exceedances Greater Than 1.0

EPA Region	Metropolitan Area (CMSA/MSA)[1]	1985-1987 Design[2] Value	1985-1987 Avg[3] Est Exc	1987 2nd[3] Daily Max One-Hr	1987 Est[3] Exc
I	Boston, MA (CMSA)	0.14	2.2	0.14	4.3
I	Conn./Mass., CT-MA (Note #4)	0.17	5.8	0.17	11.6
I	*Hancock County, ME	0.13	1.3	0.12	1.1
I	*Kennebec County, ME	0.12	1.2	0.09	0
I	*Knox County, ME	0.15	4.4	0.13	6.5
I	*Lincoln County, ME	0.13	2.4	– no data–	
I	New Bedford, MA	0.14	2.4	0.12	1.0
I	Portland, ME	0.14	3.4	0.14	4.0
I	Portsmouth-Dover, NH-ME	0.13	3.2	0.13	3.2
I	Providence, RI-MA (CMSA)	0.16	6.5	0.16	7.8
I	Worcester, MA	0.13	2.1	0.11	0
I	*York County, ME	0.15	4.2	0.14	4.9
II	Atlantic City, NJ	0.14	3.4	0.14	4.0
II	*Jefferson County, NY	0.13	4.7	0.13	4.7
II	New York, NY-NJ-CT (CMSA)	0.19	7.5	0.19	19.2
III	Allentown-Bethlehem, PA-NJ	0.13	1.4	0.13	3.2
III	Baltimore, MD	0.17	7.9	0.17	11.1
III	Huntington, WV-KY-OH	0.14	3.8	0.14	5.2
III	*Kent County, DE	0.13	1.8	0.15	3.2
III	Norfolk, VA	0.13	2.0	0.13	2.0
III	Parkersburg, WV-OH	0.13	1.5	0.15	3.5
III	Philadelphia, PA-NJ-DE (CMSA)	0.16	13.6	0.18	23.2
III	Pittsburgh, PA (CMSA)	0.13	1.7	0.14	4.1
III	Richmond, VA	0.13	1.3	0.14	3.0
III	Washington, DC-MD-VA	0.15	6.2	0.16	10.5
IV	Atlanta, GA	0.17	13.5	0.17	15.0
IV	Birmingham, AL	0.15	3.2	0.14	3.1
IV	Charlotte, NC-SC	0.13	3.0	0.14	4.0
IV	Jacksonville, FL	0.16	2.1	0.12	1.1
IV	Lexington, KY	0.13	1.6	0.11	1.1
IV	Louisville, KY	0.16	4.0	0.13	2.0
IV	Memphis, TN-AR-MS	0.13	2.0	0.13	2.0
IV	Miami-Hialeah, FL (CMSA)	0.15	2.1	0.15	3.1
IV	Montgomery, AL	0.14	2.2	0.14	4.3
IV	Nashville, TN	0.14	3.2	0.14	3.2
IV	Raleigh-Durham, NC	0.13	1.4	0.13	3.2
IV	Tampa, FL	0.13	2.1	0.16	4.2
V	Chicago, IL-IN-WI (CMSA)	0.17	7.4	0.18	12.8
V	Cincinnati, OH-KY-IN	0.14	1.6	0.15	2.1
V	Cleveland, OH	0.13	1.8	0.13	2.2
V	Detroit, MI (CMSA)	0.13	2.0	0.13	2.1
V	Grand Rapids, MI	0.13	1.3	0.14	3.0
V	Indianapolis, IN	0.13	1.3	0.12	1.1

Table 1. (Continued)

EPA Region	Metropolitan Area (CMSA/MSA)[1]	1985-1987 Design[2] Value	1985-1987 Avg[3] Est Exc	1987 2nd[3] Daily Max One-Hr	1987 Est[3] Exc
V	*Kewaunee County, WI	0.13	1.9	0.14	5.8
V	Milwaukee, WI (& Sheboygan,WI)	0.17	3.7	0.20	12.9
V	Muskegon, MI	0.17	6.0	0.18	11.0
VI	Baton Rouge, LA	0.14	3.0	0.16	5.1
VI	Beaumont-Port Arthur, TX	0.13	2.1	0.13	3.2
VI	Dallas-Fort Worth, TX (CMSA)	0.16	6.1	0.14	5.2
VI	El Paso, TX	0.16	9.0	0.17	11.1
VI	Houston, TX (CMSA)	0.20	19.1	0.18	20.8
VI	*Iberville Parish, LA	0.13	2.4	0.13	2.1
VI	Tulsa, OK	0.12	1.1	0.12	1.0
VII	St. Louis, MO-IL	0.16	5.4	0.17	8.0
VIII	Salt Lake City, UT	0.15	3.8	0.11	1.0
IX	Bakersfield, CA (Note #5)	0.16	35.1	0.16	47.6
IX	Fresno, CA	0.17	30.5	0.17	42.6
IX	*Kings County, CA	0.13	5.6	0.13	5.6
IX	Los Angeles, CA (CMSA)	0.35	143.5	0.32	141.2
IX	Modesto, CA	0.15	16.2	0.15	20.8
IX	Phoenix, AZ (Note #5)	0.14	2.4	0.11	0
IX	Sacramento, CA (Note #5)	0.17	9.7	0.17	14.6
IX	San Diego, CA	0.18	12.5	0.18	26.8
IX	San Francisco, CA (CMSA)	0.14	3.4	0.15	4.1
IX	Santa Barbara, CA	0.14	1.7	0.13	3.4
IX	Stockton, CA (Note #5)	0.14	8.1	0.12	(inc.)
IX	Visalia, CA (Note #5)	0.15	11.9	0.15	21.6
X	Portland, OR-WA (CMSA)	0.15	1.8	0.11	1.2

* Not a metropolitan statistical area.

1. Metropolitan Statistical Areas are defined by the Office of Management and Budget, and include a central county and adjacent counties, if any, which interact with the urban area.
2. The air quality design value is the fourth highest monitored value with three complete years of data since the standard allows one exceedance for each year. This value may differ from the actual State Implementation Plan control strategy value due to air quality modeling considerations such as the level of transported ozone.
3. The National Ambient Air Quality Standard for ozone is 0.12 parts per million (ppm) daily maximum one-hour average not to be exceeded more than once per year on average. The average estimated number of exceedances column shows the number of days the 0.12 ppm standard was exceeded on average at the site recording the highest design value after adjustment for incomplete, or missing days, during the three year period, 1985–87. The highest design value and the highest estimated exceedances for just 1987 are shown in the last two columns. These two values may be from two different monitoring sites.
4. Connecticut – Massachusetts includes Bristol, Hartford, Middletown, New Britain, New Haven, and New London, CT and Springfield, MA MSAs.
5. Incomplete data at this time, thus expected exceedance estimate is preliminary. However, the air quality status with respect to the standard will not change.

Source: U.S. EPA, "Environmental News," Press Release, May 3, 1988.

would be required in a city with non-attainment levels such as those in Cincinnati; the "medium" package of up to 50 percent is for cities such as St. Louis; and all or part of the "high" package would be needed in about 20 cities. While the EPA staff developed a representative program for attainment in 5 years, the measures specified for other than minimally impacted areas are so Draconian as to defy responsible discussion. For example, attainment in five years in the severe areas would mean a 50 percent reduction in driving and immediate closure of a number of industries, among other expensive and intrusive measures. Strikingly, though, the measures suggested for attainment if the "low" areas were given 7, the "medium" areas 12, and the "high" areas 22 years are almost as severe. For the "high" areas, still required are up to 40 percent reduction of vehicle miles traveled *and* conversion of 50 percent of fleets to methanol *and* relocation of industry, among other difficult measures.

The story that this analysis tells is that for the most severe areas, numerous actions on all fronts are required to achieve attainment. A non-trivial example of the "New Point Source Controls" category is bakeries; the ethanol formed when dough rises and is baked would be collected and incinerated.[16] Up to 174 industrial source categories with emissions greater than 1,000 tons per year are to be controlled more tightly to get a 6 percent reduction in VOCs by revisiting/tightening existing regulations. Innumerable consumer products are to be reformulated or banned – ranging from spray paints and varnishes to personal care products – to achieve a reduction of 7 percent.[17] "Restrictive New Source Review" (NSR) would mean that if non-trivial amounts of VOCs were emitted, expansion or renovation of existing facilities or introduction of new plants would largely be precluded. The representative menu of transportation control measures selected for this long term program (which gets 7.8 percent VOC reduction) include a gasoline tax of several dollars a gallon, a second car use-ownership tax of $1,000 or more per year, mass transit (which gets 0.4 percent), a tripling of parking fees, 20 percent of workers switched to a 4–day work week, and mandatory alternate driving days for private automobiles.[18]

The Office of Technology Assessment in a more recent report approached the matter differently by summing up the emission reductions that could be anticipated from "source-specific control strategies currently being considered by Congress and EPA."[19] The OTA authors stress that they were able to analyze only three-fourths of the known hydrocarbon inventory, and more reductions might be possible. They explicitly did not consider transportation control measures. They applied the expected reductions in 1993, 1998, and 2003 from the measures analyzed to conditions in each of the non-attainment areas, and concluded that while some cities would more than attain and others would be close by 1993, "For most cities ... projected reductions fall considerably below the amount needed to meet the standard."[20] It is also striking that though the measures analyzed typically brought greater reductions per unit source in the out-years, the competing influence of population growth meant that net reductions from the 1985 base would be essentially flat from 1993 to 2003.[21]

Several important messages can be drawn from the EPA scoping effort and the OTA study even though in detail their results may only be illustrative. First of all, there are no "big fixes." That is, there are no single or few sources of hydrocarbons which, if controlled, would bring attainment. Second, the new control options bear to a large degree on consumers and small commercial and industrial sources, not on large stationary sources which are already subject to stringent controls in non-attainment areas. Third, most of these changes will take substantial time and capital expenditure to effect; processes will have to be modified, formulations changed, and equipment ordered, built and installed. Finally, even taken together, and with optimistic estimates of actual performance in the field, the control options realistically do not add up to enough to bring attainment of the current standard for the most severely affected areas, even well into the next century.

Put differently, these measures or others with similar effect would have several implications for a resolute drive to attainment in the most severely affected areas. First, current transportation patterns *and* the vehicles and the fuel they use would have to be transformed. Second, the industrial structure would have to be changed and possibly shrunk. Third, also needed would be a wise variety of small "fixes" to business operations, products, and consumer behavior that would be more expensive and intrusive than those ever contemplated before. And finally, population growth would likely decline or population actually shrink as some industrial jobs disappeared, increased costs of doing business dampened other activity and the cost of living rose; this shrinkage would feed back to further reduce emissions. And *still* attainment of the ozone standard as now embodied in the CAA could prove an unobtainable goal.

These conclusions must be tempered by the fact that ozone formation remains but poorly understood, and some of the research findings now surfacing may lead to control strategies that can be more effective than brute-force reductions in hydrocarbon emissions. It is further true that new emission sources may be discovered (such as the operating vehicle emissions hypothesis indicated in note 14) that are amenable to technically and economically feasible reduction. Further, in the longer run new technologies and now scarcely-conceivable shifts in ways of living may bring substantial reductions in the problem. On the other hand, even the measures identified above will be extraordinarily hard to adopt, implement and enforce. This is because they affect myriad activities and strongly felt personal and regional interests and entail substantial expense and personal and social disruption. It appears that even with a very large effort ozone levels in some places will remain above the standard as found in the CAA into the indefinite future.

The Prevailing Response and Some Implications

Continuing non-attainment of the ozone standard has led EPA to propose a new strategy under the existing CAA and brought a number of legislative proposals. While there are substantial differences in detail, these proposals all build on the underlying premise of the CAA: That all areas must be brought into attainment of the health-based standard as currently defined, and soon.

In each of these proposals the deadlines are extended, and in all but the EPA strategy definite attainment dates are given. (EPA allows the most serious problem areas to determine their own target date, but requires a minimum of 3 percent/year improvement in addition to the improvements brought by existing measures and new Federal-level controls.) Regions are characterized by the seriousness of non-attainment and somewhat different treatment and deadlines are typically provided for. New Federal controls covering all of the nation are imposed, and to one degree or another regions are required to adopt specific control measures identified at the Federal level. In addition, local areas are to adopt such other controls as necessary to attain by the specified date. Federal sanctions of the sort now in the CAA are imposed for failure to submit and implement a plan for attainment in all of the proposals, with somewhat different implementation of these sanctions for failing actually to meet the standard. Some of the proposals have special provisions relating to interstate transport of ozone and its precursors. One, S.1894, also requires EPA to establish an additional longer term ozone health-based primary standard and a secondary standard to protect against non-health effects, unless EPA determines that these new standards are not appropriate.

Several premises underlie these proposals. The first, of which more will be said in the next section, is that attainment of the standard everywhere in a reasonable time is a desireable absolute end that must be achieved by Federal action and by Federal requirements imposed on the states. The second is that the current form and definition of attainment of the one-hour standard is appropriate. The third is that the control approach of the past is essentially correct, and that what is needed is an extension of restrictions on hydrocarbon emissions (and perhaps NOX emissions). Implicitly, past failure to attain is seen as a result of failure to implement and enforce sufficient measures. The fourth, is that attainment is physically and technically possible, and in the next decade or at most two. The fifth, is that pressing as hard as possible and setting tight deadlines is the way to secure maximum progress. The last is that enough is now known about the physical origins of ozone and potential control measures to define a course of action on which the nation should commit itself for the decades to come.

Concerns can be raised about each of these premises, though a full discussion of them is not appropriate here. Some examples of such concerns: The definition of "attainment" under the current approach may not be fully protective, and "non-attainment" may not impose significant health risks because of the idiosyncratic way in which regions are determined and ozone levels monitored. The focus on peak levels may not prove to be the most effective way to reduce health risk, depending on local conditions. Scientific understanding of the atmospheric chemistry is incomplete, and recent findings suggest that the processes are more complex than previously thought. Indeed, some measures involved in a brute-force reduction of precursor emissions everywhere may exacerbate rather than improve the problem. The hydrocarbon emissions inventory may be incomplete, and measures directed toward other sources may be more productive. Some regions may be such that no discernible control measures will bring attainment, however stringently they may be defined, implemented and enforced. Finally, information about what constitutes an ozone level at which adverse effects are found is being refined, and with new understanding, the design of the standard itself may prove to be faulty – possibly "overprotecting" in some circumstances and "underprotecting" in others.

More basic, however, is concern about the dynamic effect of pressing as hard as possible and setting tight deadlines. Briefly put, to the extent that the standard proves politically unattainable (because local measures prove too Draconian to impose), they set the government on a course of failure which has the negative consequence of lowering respect for basic institutions. When measures impose costs beyond those which persons affected think remotely reasonable, violations are encouraged which promote scofflaw attitudes toward all environmental protection measures. Similarly, with such measures the incentives for avoidance and evasion become so great that enforcement can become troublesome: self-policing falls down; implicit collusion among enforcers and the regulated community can arise when strict implementation would cause, for example, large scale local job loss.

The record under the existing CAA is instructive here: regions in violation have not imposed available measures that would have brought ozone levels down; anecdotal evidence suggests that local authorities have not acted with equal vigor against all pollution sources; and citizens as a whole have strongly resisted measures such as strict enforcement of vehicle emissions limits through Inspection and Maintenance programs. Congress has refused to allow EPA to impose unpopular measures such as those which restrict driving. And when deadlines have come, they have been cynically extended, rather than have sanctions imposed. Through it all, citizens have not been "voting with their feet." Regions out of attainment have continued to attract new residents, sometimes in large numbers such as in Southern California where ozone levels are the worst in the country.

There is another set of potential consequences as well. Pursuing this course creates incentives to avoid discovering or recognizing the problem in the first place. This is because the regulatory response to the discovery of adverse effects can not be adjusted to their severity, or to the costs of control. For example, local officials will be loathe to locate new monitors where readings may be at peak levels. Or again, Federal officials who know that tightening the standard would bring many more areas of the nation into non-attainment, lead to future non-attainment for others, force disruptive measures into being, and set the environmental enterprise up for failure may resist new scientific evidence that suggests such tightening is appropriate. That means some who could be protected at reasonable cost will be left at risk. In either case, a process by which action could be graduated to the severity of the problem would change the incentives. In short, by tying the risk management *action* to the risk assessment *result*, incentives are created which may lead to less protection as a whole.

In general, it can be argued that the programs under consideration can have unintended negative consequences. They may serve to lock in less effective and more expensive control strategies. They may cause both over control and under protection. They may lead to distrust of, and loss of confidence in, environmental protection as a whole. They may lead to an uneven and unfair enforcement of environmental regulations. And they can have negative effects on the operation of, and respect for, government as a whole. These potential consequences, however, are not a part of the debate among the different proposals now being considered. Again, one reason is that they do not comport with the basic premise of the CAA: all citizens must be provided with air free of harmful ozone levels.

Ozone Reduction: How Much, How Soon, and at What Cost?

The discussion above points out ways ozone can be reduced and the limitations to what can be done even with very strong measures. It also suggests some of the implications of a response that simply tries harder to do more of the same sorts of things that have been done before. What is missing in the current debate, however, is consideration of *how many* of the control actions it makes sense to take, *how soon*, *where*, and *under what circumstances*. As noted, these questions are irrelevant as the CAA was conceived and now operates.

In conception, the CAA reflected the view that freedom from harmful air pollution was akin to a fundamental right that government should guarantee to all its citizens. The difficulty of making good on that guarantee was underestimated, but the response has been to put off the due date, not to question the goal or the basic approach adopted to meeting it. In operation, the purpose of the Act and its binary structure of in- or out-of-attainment does not comport with any consideration of degree, whether of actual damages caused by ozone or of the difficulty in making further progress against it. It also precludes consideration of any factor other than health, and that in a restricted way – on a pollutant by pollutant basis.

There is another way of looking at ozone control and that is in terms of the net effects on all the values government traditionally fosters. These include, among others, overall human health, ecological security, and satisfaction of social and economic wants. This approach suggests that ozone control actions must pass a balancing test, and be set in a broader context.

The beginning point for such an approach is to estimate the health and ecological consequences of reducing ozone levels. The categories of adverse effects comprehended under the ambient air quality sections of the CAA were described above. In fact, though, these categories are not complete. For example, reduction of hydrocarbon emissions can also avoid human exposure to chemicals with other adverse health consequences. Some of the VOCs that would be controlled are known, probable, or possible carcinogens which, while not producing sufficient risk to justify regulation under other provisions of the CAA, nonetheless would offer benefits from reduction as an ancillary outcome of further control. Further, while these sections of the CAA appear to posit zero health benefits for reductions below the primary standard, it is probable that thresholds for harm are lower for some individuals, and health effects would be avoided if ozone were below levels defined by the Act as "safe." Moreover, evidence is mounting for adverse health effects when exposure to levels of ozone below the standard occur over longer periods or are repeated. Hence, health benefits would quite possibly occur with added controls in areas where the one-hour standard would lead to none. The non-health consequences of elevated ozone levels for materials, ornamental plants, forests, crop yields, and visibility are also documented. While the CAA offers the possibility of setting a more restrictive secondary standard to protect these values, the non-health benefits from the primary standard are not part of the official decision process.

The issue, though, is *how many* of these effects would be eliminated, with *how much* of a reduction of ozone, from *what* baseline level and in *each* location. Here the evidence is much more inconclusive, partly because the structure of the CAA has offered no incentive to gather it, and partly because of the inherent difficulty in doing the epidemiological and other studies from which such evidence could be gathered. Enough studies have been conducted, however, to assure that rough estimates of such effects can be gathered.[22]

There is an obvious desire to reduce adverse effects whatever their number and individual significance. But individuals and governments are more concerned about reducing some effects than others, and in reducing a large number of effects than a small number. Any balancing approach requires some scaling of the importance of the action taken. That makes the estimation of the benefits derived desirable, preferably for different levels of action, so they can be compared if only roughly to the costs. At one level this could be done with a simple toting up of the number of health incidents potentially avoided. To this could be added qualitative statements of plausible consequences of sub-clinical physiological effects discerned.

While exposure data would only be indicative, investigations of this sort could provide a rich and useful picture of the health consequences of ozone pollution. Similar efforts could estimate the effects on non-health endpoints such as visibility and damage to crops, ornamental plants, forests, and materials. While inherent uncertainties mean that such information must be interpreted with caution, it could provide telling insights to citizens and public health officials regarding the seriousness of the ozone problem in comparison to others. Standing in bold relief is the paucity of present decision-driving data which only provide peak levels at one monitor somewhere in each air management area.

It is possible to go further and to summarize the negative effects of ozone using a common metric. Monetized losses to economic values are reasonably straightforward to calculate. Substantial progress has also been made in devising methods for ascribing monetary values to non-market effects avoided. Those could be utilized for at least some of the negative outcomes. While again such measures must be interpreted with caution, as an input to decisions they can provide an indication of the magnitudes involved.

Studies of this sort are mandated under Executive Order 12291. However, relevant provisions of the CAA prevent them from being considered by the EPA Administrator in setting the standard. Much of the CAA legislation now being considered further lessens the usefulness of such information in other EPA regulatory decisions. Most important, though, ozone controls and health and environmental consequences are local or regional in nature. Unless disaggregated data are available, and local communities have the flexibility to use them, they cannot affect decisions about the actions that create the benefits and impose the costs. The single national standard for ozone precludes such flexibility. As noted before, however, any such data are only a beginning point in evaluating the wisdom of different levels of ozone control.

There is a strong view that no citizen should be involuntarily exposed to air that might be unhealthy. Values of this sort are not unique to the environment; there is a similar sense that all children should have high quality education (and adequate pre-natal care, nutrition, etc.), that citizens should be safe from crime in their homes and on the streets, that people not be involuntarily unemployed, and so forth. Like those associated with healthy air, these values are similarly relevant to public policy decisions, and action to achieve them draws on the same social resources to accomplish. How much to accommodate *each* when it is impossible to achieve *all* (or any completely) is one of the major tasks of our political process.

In deciding how to accommodate these competing values the other side of the equation comes into play: estimating the extra social cost or risks for incremental gains in each. As with benefits, these should be reckoned as broadly and completely as possible. Also as with benefits, some aspects are more tractable to quantification than are others, but even rough estimates provide valuable inputs to the decision when a balancing approach is taken. It is also similar because these costs and risks can vary with situation and location.

At the top of some lists of the costs and risks of further ozone control is the adverse health and environmental impacts that might be *occasioned*. Each regulatory action has many and far reaching effects. For example, getting more pollution of one kind reduced in one place may mean that risk is increased elsewhere. Further, as noted above, to reduce ozone levels in some areas will likely require that some economic activity be suppressed, to be moved elsewhere in this country or overseas. While no one would suggest that environmental protection should be held hostage to threats of job loss, it is equally true that the mental distress of personal disruption and the added physical and psychic health stress of loss or reduction of incomes should not be ignored either. Neither, of course, should such gains be ignored when they result in areas where economic activity is fostered.

Too, there are only so many people and so much of our national resource of scientific and engineering talent that can be devoted to the environment. There is only so much attention local officials can spend (and citizens will bear) on environmental matters. Those resources expended to reduce ozone levels are not available elsewhere, and the subsequent environmental losses must be counted against the gains. Paradoxically, then, some measures that improve health with regard to ozone exposure may leave overall health and environmental quality worse off. The net health improvement from more ozone control is surely less than the total, but by how much is a matter of fact, and the facts have not been gathered.

The distribution of adverse health effects among people will also change. Questions of equity thus arise in respect to benefits as well as to costs of further control. As to costs, some will be spread widely, but others will be concentrated on those who bear substantial reductions in income, lowered job and other opportunities, and the personal disruption of out-migration. Just as the common view weighs serious health effects higher than even more numerous discomforts, so too are such economic exactions usually judged more serious in policy formulation.

In addition to these observable effects, the resources used to reduce ozone have to come from somewhere. The "where" they come from is in production of other goods and services consumed in the economy or invested for the future. Healthier air adds to the quality of life, but that truth does not alter the fact that the things given up can also have an effect on the quality of life. And not only for this generation. Economic growth can be lowered or restructured and international competitiveness dampened, leaving those in the future with fewer opportunities than they otherwise would have.

One partial measure of what is given up is the value of the resources expended in proposed controls, a subject studied more intensively than other aspects of this issue. These costs are difficult to summarize because they depend so heavily on the baseline assumptions used, the expected cost and effectiveness of controls in practice, and the technological developments that may or may not lead to lowering those costs. Too, there are non-market costs which are difficult to estimate and are left out of most of the calculations. For example, what is the loss in welfare to be ascribed to changed shopping or commuting patterns resulting from restrictions on automobile use? In turn, how much should be subtracted from that total because lower congestion might speed those trips that are actually taken? Again, though, some understanding of the magnitudes of direct resource cost is instructive.

One approach to the issue would be to take the future cost of controls existing as of January 1, 1988, add to that the costs of programs now under EPA consideration, and then posit a representative cost per ton for the ozone reductions needed for attainment but yet to be identified. This results in a cost for (possible) attainment of something like $23 billion per year for 2000 and beyond.

Another approach would be to do as OTA does and estimate the costs of known control measures, recognizing that they do *not* achieve attainment and indeed leave the situation about stable from 1993 forward. That approach yields added costs beyond those for controls already in place of $7.8 billion in 1993, $9.6 billion in 1998, and $11 billion in 2003 at the high end, and of about $1 billion less for each of those years on the low end.[23] Summing the increased cost of existing controls already in place as they will be implemented across the country and the identified control measures found in the OTA report suggests a social expenditure for ozone control before the turn of the century of a minimum of $100 billion and up to $150 billion, even recognizing slippage in implementation. Not accounted for are the non-market costs that would be involved. Again note, a program of this sort would not be such as to assure attainment of even the one-hour ozone standard as currently defined and structured.

The public debate over what to do about ozone has not been informed by explicit consideration of either social benefits or costs. The CAA answers the questions of "How much, how soon and at what cost?" with: "As much as necessary to meet the standard, as soon as possible, at whatever cost it takes." In contrast, the balancing approach would respond with: "As much as is socially beneficial, as soon as is practical, taking deeper social values into account." The operational differences between the two approaches are striking.

Conclusions

The discussion thus far suggests that:

- The task of controlling ozone is far more difficult and far more costly on many dimensions than envisioned when the CAA was passed and amended.
- The health effects of ozone are more complex than previously thought, and a different set of design parameters for the standard may be called for if adverse health effects are indeed to be avoided.
- The current understanding of the atmospheric processes leading to ozone control and of the sources of ozone precursors may be deficient.
- There is need to evaluate and assimilate the medical, scientific, technical and economic research that has been done, and to support that which is needed to fill in the gaps.
- The deadlines built into current legislative proposals may be as technically, economically and politically impossible to meet as were those they would supersede.
- The long term consequences of yet again setting implausibly difficult tasks can be far different than supposed: rather than encouraging maximum progress, they can slow actual progress, cause health effects to be ignored, and put broader support of environmental protection itself at risk.

- There are a complex set of values in collision around any decisions made; it is not a simple matter of whether or not one is in favor of healthy air.
- There is a different way to look at ozone control, and that is in terms of what we get compared to what we give up.

For these and other reasons, it is time for a debate that would open the CAA up to substantial modification to meet current understanding of the facts and to respond to new understanding as it becomes available. It may also be time to debate a change from a pollutant-by-pollutant health criterion to one which takes a balanced approach toward health, the environment, and other social goals. Despite the intense and understandable pressure on Congress to act, it is *not* time to choose among the proposals now being considered. An adequate information base simply does not exist.

The ozone pollution situation is serious but not critical. While the anomalous weather of the summer of 1988 will lead to higher ozone levels, regulations on the books that are taking effect and being implemented appear at least to be preventing the underlying situation from getting worse. Expected action by EPA – to mandate added controls to limit refueling and evaporative emissions from vehicles and to reduce the volatility of fuel – will help still more.

The ozone problem will not soon be resolved; it is not a crisis which can be put behind us, but a problem which will require steady, dedicated work for decades to come. These factors suggest that it is more important to set a sustainable, steady course that leads to progress than to take precipitate action. Especially is that true when the action would likely be a prelude to future failure and a source of continued controversy and dissention. Therefore it would seem responsible for the Congress to limit itself to guaranteeing continued progress along the current course and to demanding that the information needed to confront basic choices be developed expeditiously. Such a decision would require substantial political courage, but the matter is of sufficient importance to warrant its exercise.

In preparation for the future decision it is important that public debate be joined on the basic premises on which ambient air protection is to be based. The values at stake are complex. Without broad public understanding of the consequences of choices the underlying support needed to stay the course may erode. The ozone pollution problem and its possible remedies touch such deep personal, regional, and national interests as to doom any policy that lacks broad consent.

More generally, the rhetoric that speaks of a United States free of environmental risk is confronting an ability to discern smaller and smaller levels of pollutants with effects on fewer and fewer people. At the same time, doing something about the risks imposed by these pollutants is now understood to have broader consequences than before. Too, the demands for other government services are growing, and the limits on the resources available to meet them are more clear. In this context the debate on ozone can serve as a proxy for some of the value conflicts that cannot be long avoided in environmental protection as a whole.

Epilogue

This chapter has discussed the ozone pollution problem and the hard choices posed in resolving it. Its burden is that the current CAA is flawed in premise and structure – perhaps so based on the understanding that existed when it was adopted and amended, but certainly so in the context of what is known now. It is not the purpose of this paper to propose a structural "fix" for the ambient air quality problem. Nonetheless, an outline of some of the elements that could be considered may be appropriate as an Epilogue.

A new policy must address at least three issues. The first is the basis on which a standard or regulatory goal should be set. The second is the way that standard will be applied across geographic areas. The third is who is to decide.

Turning first to standard-setting, the scientific search for health effects from pollutants should continue to be encouraged, but as a National Academy of Sciences report suggests, there should be a separation between finding an effect and deciding what to do about it.[24] Some effects are more serious than others, and affect more people. But the way the law is now written, it is almost as if a cancer were equivalent to a cold, one expected cancer were indistinguishable from an epidemic, and as much social disruption, other risk and economic cost were to be imposed to avoid the one as to avoid the other. Flexibility to discriminate among adverse health effects, and to allow the broader ramifications of different stringencies of standards to be taken into account, is needed.

Further, a single standard as a requirement for action does not make practical sense in a country as diverse as the United States. It would likely leave some underprotected when the benefits of cleaner air are weighed, and others overburdened when the costs of attaining the standard are reckoned. One way out of this dilemma would be to abandon the idea of a single health standard at the Federal level and instead define a range. On one end would be a maximum allowed level of the pollutant that *fully* protects against all adverse health effects, as the current standard purports to do. On the other would be a pollution level that meets the country's considered judgment of a degree of risk beyond which no one should be exposed involuntarily. In between would be an "adequate quality" standard. This "adequate quality" standard would be set to be protective against "significant" adverse health effects. How serious the effect would be to warrant the designation as "significant" and what proportions of the susceptible populations would be protected must necessarily be left to judgment in the national political process.

This range could be built into the allowed peak ozone level itself, but it might make more sense to implement it in terms of the number of times a year a standard is exceeded, for what duration, and by how much. Health risk appears to come from a combination of the level of dose and the number and duration of exposures, so all these factors are important. A policy taking this into account could be more protective, and at lower social cost, than any standard that only considered peak levels.

A structure such as this might prove an acceptable compromise between two views of the nature of the environmental protection enterprise. On the one hand it recognizes that something akin to a right exists for all citizens to be protected against intolerably high involuntary risks imposed by the collective actions of others, no matter how costly or inconvenient that protection might be. On the other hand, it also recognizes that there are other goals in a society, and that achieving them sometimes leads to greater burdens – including environmental risks – being imposed on some. In these circumstances a broader balance is struck among a series of competing "goods," based on the collective political judgment of the positive and negative consequences of alternative courses of action. Those negatively affected can act to protect themselves (for example, by avoiding exposure), can be compensated in some other way, or, also likely, unfortunately may bear an unequal share of this social burden. While this outcome is stark when it comes to health risks, however disguised, it is also an inevitable consequence of all collective decisions.

Turning back to implementing the control system, at the end of the spectrum where the risks imposed are simply not acceptable, stringent efforts should be demanded, with their specific design left to the affected communities (and those from which the problem is imported) which can choose the fairest, most efficient actions. Time will be required for these efforts to work, but firm, tight schedules with specific check points along the way are needed, together with stringent sanctions to enforce them from the Federal level. The national interest in assuring basic protection for all citizens requires no less.

That leaves those areas that fall between meeting the fully protective standard and the unacceptable level. Here is where local conditions merit the most local consideration. Each community will certainly want to move toward achieving completely healthy air, but at the same time may find itself trapped between doing better for the air and sacrificing other values its citizens also hold dear. Where to draw the line on pollution is not an easy decision, obviously, but it is one those most affected can reasonably be expected to make, and one they can reasonably expect others to honor. From those exceeding the "adequate quality" level strong measures would be expected, along with a schedule that showed progress toward attaining it. How long it would take to reach that level depends on what is feasible and what other goals would be sacrificed, and best can be determined locally.

This sketch passes quickly over many thorny problems. Among the most important is the political problem posed by ozone transport across jurisdictional boundaries that do not match air quality zones. There appears to be no ready accommodation between the dual goals of allowing those closest to the situation to use existing political structures to make the trade-offs and of preventing citizens of one jurisdiction from fouling the air for those of another. This issue will require careful attention, and perhaps require new institutional arrangements.

The approach outlined here follows from several predicates which themselves deserve debate. The first is that within limits the costs of the sort incurred in improving the environment should be commensurate with the benefits received, and that both should be counted up as holistically as possible. The converse underpins the ambient air quality sections of the CAA where, taken literally, other values are irrelevant when it is necessary to sacrifice them to avoid specific adverse health effects. The second predicate is a variant of the first. It holds that different circumstances should allow somewhat different balances to be struck. Again, this is counter to the CAA premise that every location should attain the same protective air standard, and no less. The third principle is that the people most affected should have the major, but not the only, say in the quality of their air and how it is attained. In contrast, the CAA puts the scientific discovery of adverse health effects in the position of determining the air quality that would be met. It is on these deeper grounds that fruitful debate is required.[25]

Notes

1. This paper complements but also partially overlaps "Tropospheric Ozone and Vehicular Emissions," (ORNL/TM 10908, available from NTIS) which was the product of the same research project. Also published from this project was "Ozone Pollution: The Hard Choices," *Science*, Vol. 241, September 9, 1988 and "Ozone Pollution: No Easy Choices," in the Spring, 1990 issue of *The Forum for Applied Research and Public Policy* which includes excerpts from this chapter. The research was supported by institutional funds from Oak Ridge National Laboratory and the Energy, Environment and Resources Center, The University of Tennessee, Knoxville. This research was completed in August, 1988 and this chapter does not reflect subsequent developments or new information.
2. Sections 108 and 109 of the Clean Air Act establish the process by which standards are set for ozone, one of the six "criteria" pollutants (42 U.S.C. 1857 et seq). Ground level tropospheric ozone, the "bad" discussed here, is not to be confused with stratospheric ozone which shields the planet from damaging ultra-violet rays. For 1987 areas out of ozone attainment see. U.S. EPA, "Environmental News," Press Release for May 4, 1988.
3. P.L. 100–202.
4. These include S–1894 (Mitchell), H.R. 3054 (Waxman), "Group of Nine" discussion draft, "Fields Amendment to H.R. 3054;" "STAPPA/ALAPCO Strategy," and the "EPA Post-1987 Strategy."
5. See David E. Gushee, "Ozone/Carbon Monoxide Nonattainment: Is It What It Seems To Be?" Congressional Research Service Report for Congress 88–1485, February 18, 1988, for a discussion of exposure profiles by time of day, location in the airshed, and individual activity.
6. Strategies and Air Standards Division, Office of Air Quality Planning and Standards, U.S. EPA, "Review of the National Ambient Air Quality Standards for Ozone, Preliminary Assessment of Scientific and Technical Information: OAQPS Draft Staff Paper," pp. IV–4 through IV–5 (cited hereafter as "Draft Staff Paper").
7. U.S. EPA has prepared a massive "criteria" document summarizing and interpreting the literature on ozone: Environmental Criteria and Assessment Office, Office of Health and Environmental Assessment, U.S. EPA, "Air Quality Criteria for Ozone and Other Photochemical Oxidants," (August, 1986). It has summarized and interpreted this literature in the "Draft Staff Paper." Because of the limited purpose of this essay, no effort is made here to cite the original research or to reference the further work that has been reported. It should be noted that the scientific opinion is not unanimous on these findings and that credible arguments can be made that the effects have been misspecified, or, because of the design of experiments and studies, exaggerated or understated.

8. "Draft Staff Paper," p. VIII–17.
9. "Draft Staff Paper," Chapter X and other sources. The CAA directs that "primary" standards be set and achieved to protect human health with an adequate margin of safety, but that "secondary" standards (which may be more restrictive than the primary standard) be set to protect other values. For the secondary standard, much more flexibility exists to take into account the magnitude of the prospective damage avoided. As interpreted, the requirements of a secondary standard do not have the absolutist character of the primary standard.
10. Moreover, the number of monitors is severely limited. Even large areas with known ozone problems have a very small number; there were only nine in Harris County, Texas (Houston) and 11 in the Washington, D.C. metropolitan area in 1981–1985, while hundreds or thousands would be required for a one-half mile grid to characterize actual conditions. Gushee, pp. 3–4, and American Petroleum Institute, "Ozone Concentration Data," May, 1987.
11. U.S. EPA, "Note to Correspondents," August 27, 1987.
12. U.S. EPA, "Environmental News," Press Release for May 3, 1988.
13. U.S. EPA *Progress in the Prevention and Control of Air Pollution in 1986*, January, 1988, pp. 2–3.
14. An intriguing speculation surfaced in the summer of 1988. This hypothesis is that routine (mostly evaporative) hydrocarbon emissions from operating vehicles have always been higher than supposed, adding to the base inventory, and that those emissions have been greatly increased over the past few years due to the higher volatility of gasoline in use. (In turn, that increase in volatility has been caused by the subsidized increased use of ethanol as a fuel extender, and by the increase in lighter hydrocarbons, including alcohols, blended in gasoline to meet octane requirements previously served by now-restricted lead additives.) If it turns out to be confirmed, this hypothesis will yield important new directions for ozone control strategies and may make the ozone problem more tractable, at lower cost, than previously estimated. It demonstrates the changing nature of basic understanding of the ozone situation. "EPA Staff Finds Unstudied Auto Emission Source May Thwart Ozone Attainment," *Inside EPA*, August 25, 1988, pp. 1, 9–10.
15. Office of Air Quality Planning and Standards, U.S. EPA, "Implications of Federal Implementation Plans (FIPs) for Post–1987 Ozone Non-Attainment Areas," (Draft), March 1987, Table V–2, p. V–7. (Cited hereafter as "FIP Study.") This study was a "scoping" effort which was not, and is not purported to be, a definitive analysis of ozone conditions and what may be done about them. The focus was exclusively on hydrocarbon emission reduction and did not take into account strategies which might be developed based on further understanding of atmospheric processes. It should be interpreted as a thoughtful effort by knowledgeable EPA professionals under severe time constraints that gives some understanding of the gross magnitude of the problem. [Editor's note: The plan for attainment issued by the Southern California Air Pollution Control District in 1989 for the Los Angeles Basin adopted somewhat different measures but in essence confirmed the stringency and pervasiveness of the controls suggested by EPA and the time required for attainment.]
16. "FIP Study," pp. V–61 through V–62.
17. "FIP Study," pp. V–64 through V–70; V–81 through V–82.
18. "FIP Study," p. V–98. Again, these sample programs would be modified for specific locations and are not a forecast of measures that would be adopted. Also, some observers believe that substitution of natural gas for other fuels, and new technologies such as methanol fueled vehicles could penetrate faster than expected here. For a further discussion of vehicle emissions see also the author's companion paper "Tropospheric Ozone and Vehicular Emissions" cited in note 1.
19. Friedman, Robert A., et al., "Urban Ozone and the Clean Air Act: Problems and Proposals for Change," Staff Paper, Office of Technology Assessment, April, 1988, p. 79.
20. *Ibid.*, p. 100. See also Table 3–10 of the OTA Staff Paper which shows that in 1993, after implementing all of these control measures, the best estimate would be that additional reductions from 1985 levels of about four percent would be required in the least severe non-

attainment areas, and up to 64 percent in the worst. From other information presented, these numbers would not change much for 1998 or 2003.

21. *Ibid.*, p. 100 and Tables 3–5, 3–6, and 3–7.
22. It is not appropriate here to attempt to summarize the quantitative literature on ozone health and welfare effects. Illustrative, however, are the results of literature surveys and work such as that found in Krupnick, Alan J., "Benefit Estimation and Environmental Policy: Setting the NAAQS for Photochemical Oxidants," Discussion Paper EQ 87–05, Resources for the Future, December 1986; and Kopp, Raymond J., and Alan J. Krupnick, "Agriculture Policy and the Benefits of Ozone Control," *American Journal of Agricultural Economics*, Vol. 69, No. 5 (December 1987), and subsequent unpublished work. See also Hayes et al. (Systems Application, Inc.), "Assessment of Lung Function and Symptom Health Risks Associated with Attainment of Alternative Ozone NAAQS," September 18, 1987 which summarizes some of this evidence.
23. Friedman et al., op. cit., note 19, pp. 106–115 and Table 3–11.
24. National Academy of Sciences, *Risk Assessment in the Federal Government: Managing the Process*, Washington, D. C., 1983.
25. I wish to acknowledge discussions or comments on earlier drafts of this and related papers by a number of persons which contributed to my understanding of the issues surrounding tropospheric ozone pollution. These persons include: David J. Bardin, David Bates, John M. Campbell, Jr., E. William Colglazier, Mary English, Robert A. Friedman, Bernard Goldstein, Judi Greenwald, Charles R. Kerley, Alan Krupnick, Morton Lippman, Paul Portney, Robert SanGeorge, Daniel Sperling, and five anonymous referees. A number of EPA staff members kindly supplied information and factual review of earlier drafts of work from which this paper was drawn. These include: Allen Basala, John Calcagni, Eugene Durman, Gerald Emison, Michael Jones, Bruce Jordan, David McKee, and Harvey Richmond. Obviously none of these persons bear any responsibility for remaining errors and they should not be associated with the interpretations and conclusions I have drawn.

Chapter 11

Health and Community Issues Involved in Setting Standards

MICHAEL D. LEBOWITZ

This chapter offers a discussion of air quality standards under the Clean Air Act. The major objective is to address the ways in which primary standard setting is based on adverse health effects. Other objectives are to address the importance of using standards for short-term and long-term solutions, the issues of compliance, trade-offs and costs. Within the social or public policy framework, it is also important to attempt to identify the parties responsible for pollution, in order to prevent adverse effects through communication and public debate.

Public health practices, operating within the current legal framework, give priority to preventive approaches, which are to be derived from an understanding of the causes of adverse health effects (Hill, 1965; World Health Organization, 1982b; 1983). The elements of this preventive approach can be viewed as a sequence of steps (World Health Organization, 1982b): normally expert concern leads first to descriptive studies; generic analysis of these studies then lead to more specific exposure and health effects studies; theoretically, impact estimates are then derived from such studies, and result in generic corrective measures.

A large part of the practical problem in taking a preventive approach, or promoting compliance as one of the remedial steps, derives from policy that ignores these steps. Sometimes, public concern leads not to studies of the problem, but either to inappropriate monitoring and risk assessment, or directly to incorrect (or insufficient) attempts at reducing exposure. (Sometimes, it leads only to cover-up information being provided disguised as expert opinion.) More often than not, impact statements are not derived from exposure-health effects studies.

The results of scientific studies can be utilized in modeling risk and then cautiously extending the estimates to risk assessment (Omenn, 1987), the current method of estimating impact. One does this by obtaining exposure-response relationships in appropriate populations. This may have to include determining time-activity profiles in those populations. One then determines the exposure and time-activity scenarios for the populations of concern in order to identify how many people will display specific health effects during a

M. Waterstone (ed.), Risk and Society: The Interaction of Science, Technology and Public Policy, 165–171.

particular time period. One can then determine the extent of the effect (as deaths, or in time allotments, disability days, health care person-days, etc.). These effects can be multiplied to get population impacts.

On the other hand, monitoring per se in order to estimate exposure which is then incorporated into current risk assessment models is rarely appropriate. This is because current models usually are based on crude estimates in humans or effects in animals, do not account for synergistic factors and do not control for confounding factors. On the other hand, risk assessment with knowledge gained from exposure-effects studies should lead to more realistic risk management scenarios. Building upon these scenarios one can determine the options that are available for control and thus what policies to pursue to minimize the risk to populations. Thus, it is better to have (or do) the proper exposure-effects studies to utilize in population impact assessment which can then generate appropriate corrective measures. It is also much easier to communicate such results within public debates.

Assessing the adverse health effects of air pollution in specific communities, especially for sensitive subjects, requires difficult and costly studies (Cassell and Lebowitz, 1976; Shy et al., 1978; Zagraniski et al., 1979; World Health Organization, 1982a; 1982b; 1983; Leaderer et al., 1986; Quakenboss et al., 1989). Especially complex is the evaluation of the personal and social behavior involved in the production of pollution and how this pollution produces health responses (World Health Organization, 1982a; 1982b; 1983; 1986). It is a more difficult task as there are a wide variety of contaminants which have varying temporal-spatial concentrations. Further, these contaminants may interact in producing the health effects of concern. The control task is also difficult because these pollutants are generated by different sources and, therefore, a variety of responsible parties.

One must consider also the contribution from other environments, especially indoor environments (National Academy of Sciences, 1981). These environments are thought to play a major role in the exposure to many contaminants. (Individuals spend 65–95 percent of their time indoors.) However, the nature of both the contamination and the exposure is a function of individual behavior (which is difficult to change), and such areas are off limits to most regulatory control measures. Nevertheless, the contributions of such exposures to total exposure is critical in determining appropriate policy and control. Effective assessments of this component of total exposures and responses will require extensive study, especially because the overall health impact attributable to such indoor exposures is estimated to be significant (Shy et al., 1978; World Health Organization, 1983; 1986). Therefore, short- and long-term corrective policies must reflect these additional complexities.

For example, one can evaluate the exposure and health impact of particles, specifically particles that are inhaled (PM–10). National Ambient Air Quality Standards (NAAQS), recently revised, for this size particle, are based on acute and chronic health effects from ambient exposures, in the types of studies indicated above as desireable. Population impact from estimated ambient exposure is very large because this type of air pollutant is ubiquitous, and because specific sub-populations are at higher risk. However, the impacts of work exposure or indoor exposure to PM–10 were not included. Therefore, it is not possible to determine if total exposure is equivalent to the ambient estimate as indoor/outdoor ratios can be less than, or greater than, one, depending on location, ventilation, and sources. Indoor PM is of a different chemical nature usually than outdoor PM (as the sources differ), so the health effects usually associated with each exposure type may also vary. However, the adverse impacts are likely to be more significant from indoor PM (e.g., tobacco and wood smoke), so the PM NAAQS is conservative (Cassell and Lebowitz, 1976; Shy et al., 1978; Zagraniski et al., 1979; World Health Organization, 1982a; Quackenboss et al., 1989). Control measures are also quite different for each type. Further, abatement of ambient PM will not affect exposure to indoor (or workplace) PM (World Health Organization, 1982a).

Finally there is the problem, indicated in the Criteria Document (World Health Organization, 1983), that particulates may interact with gases (such as sulfur dioxide) to produce health effects of concern. In this case, it was deemed that the interaction was not critical for chronic health effects. On the other hand, we know that indoor PM interacts with indoor gases (such as formaldehyde) (Leaderer et al., 1986; Quackenboss et al., 1989) to produce health effects of concern. Therefore, it is only appropriate studies, followed by education that will lead to policies for controlling indoor PM (and synergistic gaseous pollution) and minimize the health risk from PM (National Academy of Sciences, 1981; World Health Organization, 1986).

Once the health implications for exposed populations are determined, cost-benefit analysis can become part of the evaluation of priorities for setting air quality policy based on health. (Cost and benefit would include the population impact endpoints mentioned above.) This is not to say that health-based air quality standards should consider cost-benefit comparisons. However, control policy and laws could. It goes without saying that potential control strategies also require extensive assessment and testing, and then, most importantly, commitment to corrective measures. This is one stage at which public debate is essential.

To assess pollutants within a framework of science and policy, a WHO/EURO working group recently evaluated current levels of knowledge about air pollutants from different sources, and methods of control (World Health Organization, 1986). The judgment was that there may be adequate knowledge about most sources, and about many of the health effects of concern. On the other hand, the level of knowledge concerning the population distributions of exposure for these contaminants is lower, as is knowledge about synergism. Discussions of common control methods led to the conclusion that there were four: regulatory, technological, social, and medical. The regulatory approach, as discussed for PM^{-10}, is limited. Most importantly, because of the sources of pollution, the document indicated the need for exploring the social aspects of control in more detail. In other words, a large amount of pollution is produced by individuals and their activities. Social adjustments include personal and social behavioral factors, communication and education, cooperation and compliance.

Examples of Health Effects Studies and Outcomes

Carbon monoxide (CO) comes directly, and almost entirely from vehicle tailpipes. Tropospheric ozone, to which people are exposed, is a photochemical product of auto exhaust (nitrogen dioxide and hydrocarbons in sunlight). The more auto exhaust and sunlight, the more carbon monoxide and ozone in the ambient atmosphere (Environmental Protection Agency, 1986; Chapter 10, this volume). These pollutants usually have much lower concentrations indoors, so the most significant exposures are outdoors.

The chronic effects of oxidant pollutants and carbon monoxide have been studied, mostly in animals, but inconclusively. Acute effects on humans have been demonstrated in laboratory, field, and epidemiological studies (Shy et al., 1978; Environmental Protection Agency, 1986; Lippmann, 1989). However, the effects of these gases on sensitive individuals are not conclusive either. Without such knowledge, long-term standards and preventive measures are difficult to determine.

There is a necessity of evaluating personal factors, such as time and activity by location, behaviors that increase carbon monoxide and ozone (especially vehicle miles travelled - VMT), smoking (and other interactive exposures leading to adverse health effects), and socio-economic conditions, as these factors also determine risk and lead to various policy measures. The factors also require further study to determine population impacts (Shy et al., 1978; World Health Organization, 1982a; 1983) and the levels of correction/control necessary.

Specifically, the factors which increase VMT (mostly behavioral and social) will increase carbon monoxide and ozone. This will lead to greater impact in larger proportions of sensitive and non-sensitive populations. Although there are short-term solutions (e.g., freeways and oxygenated fuels for CO, restricted driving for ozone), long-term solutions must decrease VMT in vehicles with combustion engines. This will require different policies about types of vehicles used, freeways, mass transit (e.g., rail systems), closure of areas to traffic, usage taxes, etc.

Lebowitz and others (Lebowitz et al., 1987b; Holberg et al., 1987; Quackenboss et al., 1989) have shown independent and interactive effects on indoor and outdoor pollutants, allergens, and weather, on symptoms and lung function in children and in sensitive individuals. Examples of the interactive contaminant effects on health measurements, with meteorological and personal covariates, were shown. Using ozone as an example, researchers have been able to observe the interactions of ozone and both outdoor total suspended particulates (TSP) and passive smoking on normal children's lung function.

The interactive effects of ozone and TSP are synergistic (Lebowitz et al., 1987a). Passive smoking inhibits the effects of ozone (Lebowitz, 1984) as active smoking does in chamber studies (Environmental Protection Agency, 1986). Studies have also observed the interactions of outdoor ozone and TSP, and of TSP and gas stove use, on lung function in adults with chronic lung disease, after adjustment was made for all other significant environmental variables. These interactions also demonstrate the relative influence of time spent indoors vs. outdoors.

Other responses of sensitive individuals to air pollutants have been determined empirically in controlled human exposure and in other epidemiological studies (Environmental Protection Agency, 1986; Lippmann, 1989). Utilizing this information, and prevalence rates from epidemiological surveys, risk assessments have been conducted to identify which health endpoints would occur in what proportion of the population, and where these health impacts would occur in relation to the sources of the pollutants of concern (Shy et al., 1978; Omenn, 1987). The information generated by this process is then useful for formulating policy decisions. However, such risk assessment techniques are still evolving, and will require empirical validation before they are adequate for risk management policy.

Perception and Policy

Some adverse health effects of personal behavior (such as active smoking) were clearly demonstrated, and accepted by various governments in the early 1960's. They were followed up quickly by public health attempts at control. Usually, control activities were restricted to education, with later regulations. In general, they were not successful. However, an increased interest in personal health led to major shifts in some lifestyles in the U.S. In other countries, peer pressure among physicians about smoking and drinking led to major changes that were then transferred to the public, along with major increases in taxes on tobacco and alcohol. These policies have had reasonable success. One likely result has been a reduction in heart attacks.

While smoking and alcohol use have decreased among some population segments, some behavior such as driving habits, including speed and seatbelt use, as well as vehicle miles travelled have not changed, so that death and injury rates are still very high from these lifestyle factors. Further, the generation of particulates, carbon monoxide and ozone are still major problems. In addition some pollution, for example, the generation of indoor (and outdoor) pollution from wood burning (and kerosene heater use), and the increase in pollutant concentrations as a result of energy conservation measures (including "tightening" of homes), has actually increased.

In some cases, public interest and perception can outpace scientific knowledge about particular types of pollution (such as passive smoking). This scenario can be fueled by public organizations as well as vocal scientist advocates in the public arena (see Chapter 2, this volume for a discussion of this phenomenon). Perceptions have been tremendous stimuli to behavioral changes, because they involve welfare aspects, such as comfort, as well as health impacts (including the terror of cancer). For such pollution, public attitudes and perceptions can determine public policy. (Since the research findings lag behind public opinion, great pressure can be placed on scientists to operate as experts in the public and governmental arena, whether appropriate or not.)

Pollution control, containment or corrective policies usually have been generated through either legislation, government administration or popular votes; they rarely occur through voluntary or informal mechanisms. Unfortunately, the perception of need for corrective policy has not usually been accompanied by a willingness to pay for such remedies, especially not directly. Social and economic costs, in general, do not appear as yet to have been addressed sufficiently. For example, there is increasing polarization and isolation of individuals (the smokers vs. non-smokers battle), more costly public and work place ventilation, less work performed (because smokers have to go elsewhere to smoke). Although we assume that the balance probably favors the benefits, these measures (and their effects) require more study.

The time has come for a consistent, comprehensive plan for the study and possible regulation of pollutants to emerge from the government policy-making process. Agencies must translate the results of research into understandable, workable voluntary standards to guide individuals and governments. Better risk communication must occur, and must be two-way between government agencies and the public. Public debate must include short- vs. long-term solutions, trade-offs, and cost. Through communication, policy makers and the public should learn when to avoid negative trade-offs (such as the use of most alternate fuels that increase ozone and other oxidants during photo-oxidant seasons). Likewise, comprehensive planning and communication are necessary to determine which resources are available, and which can be used for different policies.

As with many public health issues, ultimately it will be the responsibility of individuals to change their behavior in ways that reduce pollution and limit exposures. In the case of ozone and related pollutants produced primarily by auto exhaust, this means changing driving habits to reduce overall VMT. Such individual changes in behavior have occurred for other factors that relate to health, such as diet, exercise, and active smoking. These are lifestyle changes that have become common after a perception of need. We are only beginning to see such change in other personal behavior, such as seat belt use and alcohol consumption (in educated, affluent sub-groups in some countries only). Personal behavior involving vehicle usage has not shown itself to be so amenable to change. This is somewhat consistent with public health knowledge, which identifies personal behavior as the most difficult set of induced risks to control. Thus, it will be part of our responsibility in our analytic activities to develop relevant advice to give the public, and public health agencies, for public policy.

Social forces should be the most pervasive and persuasive means for dealing with air pollution, although regulatory controls will undoubtedly have to be used as well. Social strategies could help control traffic and thus ozone (such as closing downtown sections, the use of mass transit, the "Olympic solution"), but regulation of VMT's will probably be necessary as well. These regulations could come about through public debate. New or different compliance strategies are needed as well. The bottom line is that people have to be involved in determining strategies, regulations, compliance, and how the limited national resources will be used.

References

Cassell, E., and M.D. Lebowitz. 1976. "The Utility of the Multiplex Variable in Understanding Causality." *Perspectives in Biological Medicine* 19(3): 338–341.

Environmental Protection Agency. 1986. *Air Quality Criteria for Ozone and Other Photo-oxidants.* Washington, D.C. RTP(NC): EPA.

Hill, A.B. 1965. "The Environment and Disease: Association or Causation?" *Proceedings of the Royal Society of Medicine* 58: 295–300.

Holberg, C.J., et al. 1987. "Multivariate Analysis of Ambient Environmental Factors and Respiratory Effects." *International Journal of Epidemiology* 16: 399–410.

Leaderer, B.P., et al. 1986. "Assessment of Exposure to Indoor Air Contaminants from Combustion Sources: Methodology and Application." *American Journal of Epidemiology* 124: 275–289.

Lebowitz, M.D. 1984. "The Effects of Passive Smoking on Pulmonary Function: A Survey." *Journal of Preventive Medicine* 13: 645–55.

Lebowitz, M.D., et al. 1987a. "The Epidemiological Importance of Intra-Individual Changes in Objective Pulmonary Responses." *European Journal of Epidemiology* 3: 390–98.

Lebowitz, M.D., et al. 1987b. "Time Series Analysis of Respiratory Responses to Indoor and Outdoor Environmental Phenomena." *Environmental Research* 43: 332–41.

Lippmann, M.V. 1989. "Health Effects of Ozone." *Journal of Air Pollution Control* 39: 672–95.

National Academy of Sciences (NAS). 1981. *Indoor Pollutants.* Washington, D.C.: National Academy Press.

Omenn, G.S. 1987. "Framework for Risk Assessment for Environmental Chemicals." *Washington Public Health* 6(1): 2.

Quackenboss, J.J., et al. 1989. "Epidemiological Study of Respiratory Responses to Indoor/Outdoor Air Quality." Environmental International.

Shy, C., et al. 1978. *Statement on the Health Effects of Air Pollution.* New York American Thoracic Society.

World Health Organization (WHO). 1982a. *Estimating Human Exposure to Air Pollutants.* WHO. Copenhagen/Geneva.

World Health Organization (WHO). 1982b. "Indoor Air Pollutants: Exposure and Health Effects." WHO/EURO Reports and Studies 78. Copenhagen.

World Health Organization (WHO). 1983. "Guidelines on Studies in Environmental Epidemiology." WHO (Environmental Health Criteria #27). Geneva.

World Health Organization (WHO). 1986. *Methodology in Indoor Air Quality Research,* WHO/EURO Report 103. WHO. Copenhagen.

Zagraniski, R.T., et al. 1979. "Ambient Sulfates, Photochemical Oxidants, and Acute Health Effects: An Epidemiological Study." *Environmental Research* 19: 306–320.

Chapter 12

The Abuses of Risk Assessment

DAVID S. BARON

Environmental advocates have traditionally viewed risk assessment and cost-benefit analysis with great skepticism and justifiably so. Such techniques have often been disingenuously used to thwart or delay sorely needed measures to protect public health and natural resources. Although risk assessment can sometimes be a useful tool in environmental decisionmaking, it is too value-blind and imperfect to drive the process.

The Imperative of Efficiency

Under a strict cost-benefit analysis, the most economically efficient approach to a given problem must prevail. But many of our most important values – the protection of human life, the preservation of scenic beauty – cannot be expressed in dollars and cents. This is not to say that cost is irrelevant, but characterizing environmental decisions in terms of "dollars versus lives" implicitly places the burden of proof on the public to "justify" the expense. Where involuntary exposure of citizens to toxics is proposed, the burden should be on the proponents to justify the exposure – not the other way around.

Consider, for example, a mining company that wishes to avoid the multimillion dollar cost of sealing its mine waste ponds so as to protect groundwater quality in a small town. Under a strict cost-benefit analysis the mine might have a case. It could well be cheaper to allow contamination of the aquifer and move the townspeople or truck in bottled water. But such a result could seriously intrude on individual rights. Few would argue that a mining company has an absolute right to oust citizens from their homes, to make a town unlivable, or to render a common resource unusable for future generations.

Risk assessment is also antithetical to an important goal of many environmental laws – technology forcing. Under technology-forcing statutes, health standards are set with little or no concern over the availability of existing technology to meet them: the whole purpose is to *force* the development of new technologies to ensure the desired pollution reductions. A good example is the Clean Air Act of 1977, which directed the automakers to achieve a 90 percent

M. Waterstone (ed.), Risk and Society: The Interaction of Science, Technology and Public Policy, 173–178.

reduction in new car emissions – a directive that forced the development of innovative control systems. Risk assessment is too constrained to produce such results. It can only work with existing technologies having known costs and benefits.[1]

Practical Pitfalls

Former EPA Administrator William Ruckelshaus reportedly once observed that we never would have tried to land a person on the moon if we had relied on risk assessment.[2] The technique is at its worst where there are major uncertainties. In such cases, significant assumptions must be made to estimate costs and benefits: and decisions are invariably delayed while more data are developed. There is nothing unique in the former: the problem is that the final cost-benefit ratio, with its appearance of technical precision, often obscures the tenuous assumptions upon which it is based. And the delay problem is particularly acute because it is always possible to argue that more study will produce more accurate results.

Unfortunately, the public is faced with myriad immediate health threats that will not wait while a "perfect" standard is being developed. On a daily basis, citizens are exposed to a barrage of hazardous substances in the environment, most of which are not limited by regulatory standards. For example, EPA has set air quality standards for only seven toxic air pollutants, despite the fact that hundreds of such pollutants present significant health threats.[3] The absence of standards means, for all practical purposes, the absence of controls on most of these pollutants.[4] Yet the setting of limits on even the most toxic gases has been delayed interminably by years of study.[5] Meanwhile, just 17 air toxics cause an estimated 2,000 excess cancers in the United States each year.[6]

The delays occasioned by risk assessment might be easier to swallow if at least some controls were adopted in the interim. But regulatory officials invariably use the pendence of such studies to justify doing nothing at all. A good example is EPA's refusal to require adoption of state plans to control airborne particulates while the particulate standard was under review between 1981 and 1987. The original standard adopted in 1971 set allowable concentrations for total suspended particulates (TSP), but in the early 1980's EPA began considering a new standard limited to small particulates (PM–10). EPA apparently thought it would be inefficient to require states to attack the TSP problem when the standard might be revised.[7] But the Agency failed to consider that many of the same control measures would be required *regardless* of which standard was ultimately adopted: for example, fugitive dust restrictions, limits on diesel emissions, controls on wood-burning, and smoke-control laws.[8] By suspending the plan adoption process for more than six years, EPA simply delayed adoption of many beneficial control measures that will have to be adopted in any event.

Yet another problem is the failure of many risk assessments to consider the synergistic effects of environmental contaminants. Ordinarily, the standard-setting process is directed at one chemical at a time, with risk estimates based on exposure of test animals to each chemical individually. But in the real world we are exposed to multiple toxins simultaneously, and interactions among them can have dramatic effects. In one test, the chemicals dioxin and N-methyl-N-nitro-N-nitrosoguanadine (MNNG) were applied to the skin of laboratory mice. When dioxin was applied alone, 0 percent of the mice developed skin cancer. When MNNG was applied alone 5 percent developed skin cancer. But when MNNG application was followed by dioxin application 79 percent of the mice developed skin tumors.[9] More familiar examples can be found in drug interactions in humans (i.e., when two incompatible drugs are taken together which can sometimes be fatal).

Of course, ignorance of synergism is not unique to risk assessments. The problem is that the results of these models are typically used without regard to such limitations. A case in point is a 1988 EPA proposal to clean up contamination of a portion of Tucson's groundwater supply by the suspected carcinogen trichlorethylene (TCE). Under the proposal, groundwater would be pumped to the surface and aerated: the TCE would thereby be "stripped" from the water and emitted into the air. Although TCE can also present a cancer risk through inhalation, EPA initially recommended against controls to reduce the air emissions because "the additional risk from untreated air emissions [was] very low (approximately 1 in 100 million)" and addition of a treatment system "which would reduce any additional risk to approximately 1 in 1 billion, would increase the cost by approximately $1.7 million."[10] At a public hearing on this proposal, one citizen asked EPA officials whether they had considered the cumulative risk of TCE inhalation to persons who had also been drinking TCE-contaminated groundwater for years. The Agency staffers conceded they had not.[11]

A related problem is the common failure of risk assessments to weigh the benefits of a given control measure beyond those of most immediate concern. For example, officials preparing plans to control carbon monoxide (CO) pollution will typically gauge the effectiveness of a given control measure solely by its projected ability to reduce CO levels.[12] But many CO reduction measures, particularly those designed to cut vehicle traffic (e.g., mass transit expansion, bikeways, carpooling), will reduce other pollutants as well, save energy, and save road maintenance costs. Likewise, reducing the use of pesticides will not only cut direct human exposures, but also can reduce the cost of food production and environmental impacts from pesticide manufacturing.[13]

Perhaps the most serious abuse of risk assessment is its use to cloak value judgments behind a veneer of technical precision. As an example of this practice, Barry Commoner cites a 1987 policy statement from the U.S. Office of Management and Budget (OMB) to the effect that cancer risks should not be judged by tests on particularly sensitive animals because such results tend to "overpredict the risk to humans." OMB contends that cancer risk estimates should instead be based on an averaging of test results from a variety of species exhibiting a spectrum of sensitivities. As Dr. Commoner points out, the decision to base a standard on the most carcinogen-sensitive test animal is not ordinarily a scientific judgement, but a moral or social one as to what part of the human population is to be protected: that is, whether to protect particularly vulnerable persons (e.g., infants and the elderly) or to only protect the "average" person. OMB, however, has disguised its approach "in the statistical language of science, with the lofty declaration that it is 'a more accurate estimate ... derived from a weighted average of all the scientifically valid, available information.'"[14]

The Example of Ozone Control

Many of the pitfalls of risk assessment are illustrated in the current debate on efforts to control ozone pollution in our urban areas. Ozone is an irritant that penetrates deep into the lungs, where it can cause direct damage to lung and airway cells.[15] The pollutant is formed from motor vehicle exhaust and evaporative emissions from fuel burning.[16] Under the 1977 Clean Air Act, states were supposed to develop plans adequate to ensure attainment of the federal health standard for ozone no later than the end of 1987.[17]

Unfortunately, many cities continue to violate the standard, prompting claims by EPA that attainment may not be possible in some areas absent imposition of "Draconian" measures. Some commentators suggest that the costs of attainment may not be worth the benefits, and that we should at the very least study the matter further before requiring additional control measures. Dr. Russell (see Chapter 10, this volume) suggests that the country needs a much more informed debate on the costs and benefits of achieving the ozone standard, and that choices on what trade-offs to make should be left to each individual community.

It is hard to argue against further study, but a delay in serious efforts to attain the standard is not justified either by scientific uncertainty or the supposed cost of compliance. The existing health standard for ozone is already a fairly weak one: according to EPA's own Clean Air Science Advisory Committee, the current standard is not fully protective of human health and does not protect the more sensitive members of the public. The Committee urged EPA to tighten the standard by one-third in 1987, but the Agency has yet to act on that recommendation.[18] Thus, albeit by omission, there has already been a balancing of public health concerns against practicalities. Secondly, many cities could in fact attain the standard by 1993 using available control strategies, such as auto emissions testing, control on emissions of volatile organic compounds, stage II vapor recovery, and tighter fuel volatility standards.[19] Most of these measures are not new, none of them are "Draconian," and most would provide benefits far beyond the reduction of ozone levels (including reduction of toxic air emissions and other pollutants). The problem is that state and local governments have consistently been unwilling to implement these or other measures for political reasons. Thus, many urban areas have refused to adopt Stage II vapor recovery, auto emissions testing programs, measures to reduce vehicle traffic, or stringent controls on solvents simply because they lack the political will.[20]

Significantly, cities that have been subjected to the threat of Clean Air Act sanctions have moved much more aggressively than their counterparts in attacking air pollution problems. In states like Arizona, Colorado, California, and Pennsylvania sanctions or the threat thereof have prompted adoption of clean burning fuels programs, auto emissions testing requirements, trip reduction programs, and other innovative measures.[21] Some of these measures were actually unknown at the time clean air standards were adopted. The legal mandate to meet a fixed, non-negotiable standard forced the development of creative solutions. A balancing approach is simply too myopic to provide such incentives.

Moreover, the idea of localized balancing of clean air benefits against costs is too insensitive to individual rights. Such an approach allows the power structure in a community to effectively tell vulnerable persons that they must either accept unhealthy air or move out of town. The whole purpose of having a national Clean Air Act that establishes a Federal right to clean air is to protect citizens against this kind of local arbitrariness, in much the same way as Federal law protects minorities from localized job discrimination, denial of voting rights, and school segregation. We do not allow localized balancing in these cases because we consider the values at stake to be too important to be subject to the whims of local prejudices. There is today the same kind of national consensus supporting a Federally-guaranteed right to clean air, as there has been throughout our history for other important civil rights.

Conclusion

Where its limitations are clearly understood, risk assessment can play a useful role in environmental decision making. Information on known health risks and the cost of control measures obviously can be helpful and important in choosing appropriate courses of action. Unfortunately, risk assessment is increasingly being held forth as some sort of high priest of environmental decision making that should be determinative of the outcome in every case. Opponents of government regulation have seized upon the technique as a convenient way of reducing complex public health questions to simple mathematical problems that can be resolved by plugging numbers into a cost-benefit formula. Such a narrow approach not only ignores the inherent flaws and limitations of risk assessments, but also sweeps aside important goals that cannot be expressed in numeric terms. Given these shortcomings, risk assessment should be viewed only as one informational tool among many that can contribute to rational and humane decision making.

Notes

1. According to some observers, environmental laws that require use only of available technology are stifling innovation and investment in alternative technologies. A. Pollack, "Innovators and Investors Hindered in the Business of Pollution Control," *New York Times*, August 29, 1989, p. 1.
2. American Bar Association Standing Committee Symposium. 1986. *Dealing with Risk: The Courts, the Agencies and the Congress. Environmental Law Reporter* 16, pp. 10186, 10224. (Remarks of David Doniger.)
3. S. Rep. No. 100–231, 100th Cong., 1st Sess., at 184–191 (1987).
4. Reports filed under the Superfund Amendments and Reauthorization Act of 1986 (SARA) confirm uncontrolled emissions of literally millions of pounds of hundreds of different toxic air pollutants each year. Natural Resources Defense Council, *A Who's Who of American Toxic Air Polluters*, June 1989.
5. S. Rep. No. 100–231, supra, note 3, at 189–200.
6. Ibid., at 185.
7. See 52 Fed. Reg. 24672-24676 (1987) advising states that revision of TSP plans would not be required while the standard was being reviewed.
8. See, e.g., EPA, *Particulate Controls for Maricopa and Pima Counties, Arizona* (1987) showing that control strategies needed to meet the PM–10 standard in Phoenix and Tucson, Arizona are essentially the same as those needed to meet the TSP standard.
9. Working Group on Synergy in Complex Mixtures, Harvard School of Public Health, "Synergy: Positive Interaction Among Chemicals in Mixtures," *Journal of Pesticide Reform* 2: 11–13, 1986.
10. United States Environmental Protection Agency, Region IX, Fact Sheet #4, Cleanup of TCE Contaminated Groundwater, Tucson Airport Area Superfund Site, Feb. 1988 at 6.
11. Personal observation by the author. EPA ultimately decided to add a treatment system for the air emissions.
12. See, e.g., Maricopa Association of Governments, *1987 Carbon Monoxide Plan for Maricopa County Urban Planning Area* quantifying benefits of various control measures solely in terms of their individualized potentials to reduce CO emissions.
13. Another good example of the importance of considering "secondary" benefits is the debate over whether to require Stage II vapor recovery systems at gas stations. Such systems, which suck up gasoline fumes while it is being pumped, have been touted primarily for their benefits in reducing the emission of organic compounds that contribute to ozone pollution. EPA has noted however, that another major benefit is the reduction of *other* emissions from gasoline, most notably benzene, that are carcinogenic in nature. 52 Fed. Reg. 31162 (1987).
14. Commoner, B., "The Hazards of Risk Assessment," *Columbia Journal of Environmental Law* 365: 371–373, 1989.
15. Office of Technology Assessment, *Urban Ozone and the Clean Air Act*, 1988, pp. 7–12.
16. Ibid., pp. 57–58.
17. 42 U.S.C. 7502.
18. See *Arizona Daily Star*, Jan. 25, 1988, p. 1A.
19. See OTA, supra, note 15, p. 100. Although the OTA report also found that these strategies would not be sufficient in many other cities, the report did not evaluate the potential impact of measures to reduce motor vehicle traffic, the largest source of ozone forming pollutants (p. 79). It is quite likely that most cities could attain with a combination of transportation control measures and the technology-based controls evaluated by OTA.
20. Ibid., pp. 127–128.
21. See, e.g., 53 Fed. Reg. 30224 (1988) (Arizona); Weisman, "L.A. Fights for Breath," *New York Times Magazine*, July 30, 1989, p. 15 (California).

Technology, Risk, and Society

An International Series in Risk Analysis

---

1. J.D. Bentkover, V.T. Covello and J. Mumpower (eds.): *Benefits Assessment*. The State of the Art. 1986 ISBN 90-277-2022-3
2. M.W. Merkhofer: *Decision Science and Social Risk Management*. A Comparative Evaluation of Cost-Benefit Analysis, Decision Analysis, and other Formal Decision-Aiding Approaches. 1987 ISBN 90-277-2275-7
3. B.B. Johnson and V.T. Covello (eds.): *The Social and Cultural Construction of Risk*. Essays on Risk Selection and Perception. 1987 ISBN 1-55608-033-6
4. R.E. Kasperson and P.J.M. Stallen (eds.): *Communicating Risks to the Public*. International Perspectives. 1990 ISBN 0-7923-0601-5
5. D.P. McCaffrey: *The Politics of Nuclear Power*. A History of the Shoreham Nuclear Power Plant. 1991 ISBN 0-7923-1035-7
6. M. Waterstone (ed.): *Risk and Society*. The Interaction of Science, Technology and Public Policy. 1991 ISBN 0-7923-1370-4

Kluwer Academic Publishers – Dordrecht / Boston / London

MIX
Papier aus verantwortungsvollen Quellen
Paper from responsible sources
FSC® C105338

If you have any concerns about our products,
you can contact us on
ProductSafety@springernature.com

In case Publisher is established outside the EU,
the EU authorized representative is:
Springer Nature Customer Service Center GmbH
Europaplatz 3, 69115 Heidelberg, Germany

Printed by Libri Plureos GmbH
in Hamburg, Germany